Resilient Partnerships for U.S. Military Satellite Communication Missions

Designing a Method to Assess the Impact of
Partnerships on Resilience

BONNIE L. TRIEZENBERG, KRISTA LANGELAND, GARY MCLEOD

Prepared for the Department of the Air Force
Approved for public release; distribution unlimited.

RAND PROJECT AIR FORCE

For more information on this publication, visit **www.rand.org/t/RRA584-1**.

About RAND

The RAND Corporation is a research organization that develops solutions to public policy challenges to help make communities throughout the world safer and more secure, healthier and more prosperous. RAND is nonprofit, nonpartisan, and committed to the public interest. To learn more about RAND, visit www.rand.org.

Research Integrity

Our mission to help improve policy and decisionmaking through research and analysis is enabled through our core values of quality and objectivity and our unwavering commitment to the highest level of integrity and ethical behavior. To help ensure our research and analysis are rigorous, objective, and nonpartisan, we subject our research publications to a robust and exacting quality-assurance process; avoid both the appearance and reality of financial and other conflicts of interest through staff training, project screening, and a policy of mandatory disclosure; and pursue transparency in our research engagements through our commitment to the open publication of our research findings and recommendations, disclosure of the source of funding of published research, and policies to ensure intellectual independence. For more information, visit www.rand.org/about/research-integrity.

RAND's publications do not necessarily reflect the opinions of its research clients and sponsors.

Published by the RAND Corporation, Santa Monica, Calif.
© 2024 RAND Corporation
RAND® is a registered trademark.

Library of Congress Cataloging-in-Publication Data is available for this publication.

ISBN: 978-1-9774-1258-4

Cover: U.S. Air Force photo by 1st Lt. Charles Rivezzo.

About This Report

The U.S. Space Force (USSF) is seeking to enhance the resilience and robustness of its space operations. Resilience assessments rely on a wide variety of inputs, from quantitative hard numbers, such as the number of satellites or the jam resistance of waveforms, to qualitative assessments regarding the impact of more-subjective factors, such as the incorporation of coalition and commercial partners into USSF missions and the subsequent impacts on organization, tactics, and training. Given the subjective nature of many of these inputs, there is an ongoing need to assess the continued relevance of these inputs over time. Therefore, the USSF tasked RAND Project AIR FORCE (PAF) with developing a methodology by which resilience criteria could be defined, assessed, applied to decisions, and evaluated *over time,* with a particular focus on the qualitative assessments of subject matter experts. To demonstrate the methodology, PAF focused on a specific mission and approach to building resilience: integrating coalition and commercial partners into the military satellite communications (MILSATCOM) mission. This report should be of interest to those seeking to understand how partnerships can be leveraged to improve the resilience of the MILSATCOM mission and those interested in methods to evaluate the validity and continued relevance of qualitative assessments about those partnerships.

The research reported here was commissioned by Space Systems Command and conducted within the Force Modernization Program of RAND Project AIR FORCE as part of a fiscal year 2020 project, "Evaluating Resilience Metrics in Technical Assessments of Alternative Space Architectures."

RAND Project AIR FORCE

RAND Project AIR FORCE (PAF), a division of the RAND Corporation, is the Department of the Air Force's (DAF's) federally funded research and development center for studies and analyses, supporting both the United States Air Force and the United States Space Force. PAF provides the DAF with independent analyses of policy alternatives affecting the development, employment, combat readiness, and support of current and future air, space, and cyber forces. Research is conducted in four programs: Strategy and Doctrine; Force Modernization and Employment; Resource Management; and Workforce, Development, and Health. The research reported here was prepared under contract FA7014-16-D-1000.

Additional information about PAF is available on our website:
www.rand.org/paf/

This report documents work originally shared with the DAF in May 2021. The draft report, dated September 2021, was reviewed by formal peer reviewers and DAF subject-matter experts.

Acknowledgments

We thank Joseph Tringe for his help in setting the objectives for this project and integrating our work into a larger effort at Space System Command, the Technical Basis Resilience Assessment working group. Our study benefitted immeasurably from conversations with various technical experts from across the MILSATCOM ecosystem, including the many members of that working group, commercial partners, and allied personnel. Additionally, we thank Jonathan Brosmer and Jonathan Cham of RAND for providing key qualitative and quantitative analytic support to the project. Without them, the authors would have little evidence upon which to base the analysis documented in this report.

Other RAND contributors include the Survey Research Group team who programmed and administered the community attitudes survey that is a core component of the methodology; Mel Eisman, whose research supported development of the partnership models used in our analyses and documented in Appendix A; Zach Haldeman, who contributed insight on alternative quantitative analyses; Elizabeth Bodine-Baron, who provided critical management oversight; and, finally, the many researchers who generously shared their time and expertise when encountered in the virtual hallways that the COVID-19 pandemic imposed upon us. We are also greatly indebted to our peer reviewers, Colby Steiner, Mathew Sargent, and Steve Flanagan, all of RAND.

While we acknowledge the contributions of all the above, we remain solely responsible for the quality, integrity, and objectivity of our assessment and recommendations.

Summary

Issue

The U.S. Space Force (USSF) is seeking to enhance the resilience and robustness of its space operations. Resilience assessments rely on a wide variety of inputs, from quantitative hard numbers, such as the number of satellites or the jam resistance of waveforms, to qualitative assessments regarding the impact of more-subjective factors, such as the incorporation of coalition and commercial partners into USSF missions and the subsequent impacts on organization, tactics, and training. Given the subjective nature of many of these inputs, there is an ongoing need to assess their continued relevance over time. This report documents our development of a methodology by which resilience criteria can be defined, assessed, applied to decisions, and evaluated *over time,* with a particular focus on the qualitative assessments of subject matter experts (SMEs).

Approach

An overview of our general methodology is provided in Figure S.1. Our research focus is on the qualitative data steps needed to generate inputs to a quantitative analysis. To demonstrate the methodology, we focused on a specific mission and approach to building resilience: integrating coalition and commercial partners into the military satellite communications (MILSATCOM) mission. We used semistructured interviews to elicit logic models regarding how and why integrating partners into MILSATCOM missions could impact resilience. Using insights from those interviews, we formulated explanatory paired logic statements about how partnerships impact resilience. We then used these paired logic statements in a community attitudes survey that was designed to ascertain whether the paired items are independent factors that should be modeled as separate inputs in later quantitative resilience modeling. By giving additional operational context to half of the survey respondents, we also sought to measure how these factors change based on operational context.

We used the results of this factors analysis to perform an exemplar quantitative analysis, but because of the confounding factors highlighted by our factors analysis, we do not draw conclusions from the quantitative analysis.

Findings Regarding the Methodology

- Factor analysis is useful in identifying independent factors—even for such highly ambiguous concepts as resilience and partnership—but factor analysis cannot provide insight into the confounding factors that create dependencies between factors.

- Consensus on a topic as ambiguously defined as resilience is deeply affected—not, as we had hypothesized, by time or operational context, but by confounding factors.
- The specificity of vignettes appears to influence SME assessments of how operational context changes the factors of resilience. A more specific vignette appears to force experts beyond preconceptions and to confront *conventional wisdom*.

Findings from Survey of Community Attitudes and Factors Analysis

Personnel surveyed overwhelmingly believe that the USSF, in responding to adversary attacks,

- has insufficient MILSATCOM resources to achieve resilience on its own (there is less consensus on whether the United States has sufficient diversity of resources)
- lacks the tools and processes needed to integrate coalition and commercial partners in MILSATCOM operations
- lacks the tools, training, and procedures necessary to rapidly reallocate MILSATCOM resources (whether alone or with partners) and recover in an operationally relevant time frame.

More-detailed analyses regarding how best to integrate coalition or commercial resources may be biased by the lack of trust that those resources can or will be properly integrated.

Based on these findings, we recommend that **the USSF should prioritize the development of tools and processes capable of reallocating MILSATCOM resources in an operationally relevant time frame and train operations personnel in their use.**

Figure S.1. Methodology Overview

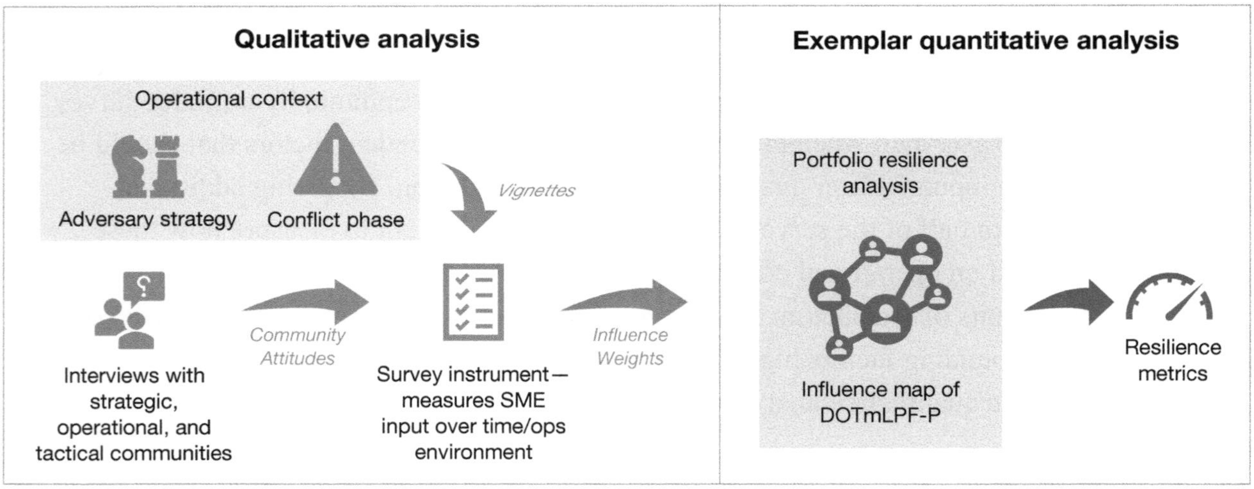

NOTE: DOTmLPF-P = doctrine, organization, training, materiel, leadership and education, personnel, facilities, and policy.

Contents

Figures

Tables

Chapter 1. Introduction

Background

Space Operations and the Need for Resilience

The U.S. military has relied on services from space since the inception of satellite technology. Initially, the reliance was focused on weather sensing and early missile warning,[1] but then increasingly on military satellite communications (MILSATCOM), and then positioning, timing, and navigation services. The resilience of these services—that is, their ability to maintain services even while under attack—has always been a consideration in the design of military systems.[2] However, as the U.S. military becomes increasingly reliant on space systems, those systems become more attractive as targets to potential adversaries. In the past decade,[3] there has been an increased focus on how best to achieve continuity of the U.S. military's space services if an adversary were to deny, disrupt, degrade, or destroy those services.

One approach to enhancing the resilience of MILSATCOM missions is to increase the use of coalition satellite communications (SATCOM) services or services from commercial companies.[4] Use of commercial or coalition SATCOM by the U.S. military is not new. Commercial SATCOM services (first seen in 1965) pre-date dedicated military satellites by about a year, and the U.S. military has always been a customer of those commercial companies.[5]

---

[1] The U.S. intelligence services' use of space for imaging and surveillance predates the military use of space services.

[2] For example, the first U.S. MILSATCOM system was the Initial Defense Communications Satellite Program (IDCSP), a fleet of small spin-stabilized spacecraft in medium earth orbit that saw its first operational use in the Vietnam War. The IDCSP constellation required 12 satellites to be fully operational, and each satellite had a six-year life. Between 1966 and 1968, 27 satellites were successfully launched (another seven were lost on launch), a quantity that provided significant operational resilience. See NASA Space Science Data Coordinated Archive, "IDSCP 3-1," webpage, National Aeronautics and Space Administration, undated; and Gunter D. Krebs, "IDCSP—DSCS-1 (NATO 1), webpage, Gunter's Space Page, undated.

[3] In response to guidance in the 2010 National Space Policy (see White House, *National Space Policy of the United States of America*, June 28, 2010).

[4] In this report, we use *MILSATCOM* to refer to the name of the mission and *SATCOM* to refer to a type of system that can be supplied by the U.S. Department of Defense (DoD), a commercial partner, or an ally to accomplish the MILSATCOM mission.

[5] The first company that the U.S. military bought satellite communications from—the Communications Satellite Corporation, or COMSAT—was not a wholly commercial endeavor. COMSAT was created by the Communications Satellite Act of 1962 as a public, federally funded corporation. COMSAT created and was the U.S. signatory to the International Telecommunications Satellite Consortium (Intelsat), an international satellite organization providing global satellite coverage, as per President Kennedy's remarks in signing the act. See John F. Kennedy, "Remarks on Signing the Communications Satellite Act, 31 August 1962," Papers of John F. Kennedy, President's Office Files, August 31, 1962.

The first U.S. military satellite constellation, the IDCSP, is also known as *NATO-1*, denoting its use by U.S. allies in the North Atlantic Treaty Organization (NATO). More recently, the U.S. military has relied heavily on commercial SATCOM for operations in Southwest Asia, especially exfiltration of unmanned aerial vehicle data for stateside exploitation.[6] In addition to using commercial SATCOM, the U.S. military negotiates international exchange arrangements for SATCOM. As of 2020, U.S. Space Command had about 20 such agreements in effect with various international partners. However, in the past few years, there have been increasing calls for more deliberate and integrated planning regarding the use of space services provided by coalition and commercial partners, including ways to assess (or measure) the resilience that such an integration would add to overall U.S. military operations.

Resilience assessments rely on a wide variety of inputs, from quantitative hard numbers, such as the number of satellites or the jam resistance of their electronic signals, to qualitative assessments regarding the impact of more-subjective factors, such as the employment of a particular organizational construct, tactic, or training program. Given the subjective nature of many of these inputs, there is an ongoing need to assess their continued relevance over time. Therefore, the U.S. Space Force (USSF) tasked RAND Project AIR FORCE (PAF) with developing a methodology by which resilience criteria could be defined, assessed, applied to decisions, and evaluated *over time,* with a particular focus on the qualitative assessments of subject-matter experts (SMEs). This report documents the methodology we developed, focusing on a specific mission and approach: integrating coalition and commercial partners into the MILSATCOM mission. The methodology, we believe, is generally applicable to evaluating factors that impact resilience for any military mission, though the specifics of the survey used to gather SME inputs would necessarily be unique to each mission.

Defining Resilience of Space Operations

The first step in achieving resilience is to define it. After several years of discussion and study, in September 2015, the Office of the Assistant Secretary of Defense for Homeland Defense and Global Security released a white paper titled *Space Domain Mission Assurance: A Resilience Taxonomy*. In the white paper, resilience is defined as "[t]he ability of an architecture to support the functions necessary for mission success with higher probability, shorter periods of reduced capability, and across a wider range of scenarios, conditions, and threats, in spite of hostile action or adverse conditions."[7] The white paper defines a taxonomy for resilience that includes six dimensions:[8]

---

[6] Keith Norton, "Commercial SATCOM Remains Vital to Military Ops," *Defense One*, August 22, 2011.

[7] Office of the Assistant Secretary of Defense for Homeland Defense and Global Security, *Space Domain Mission Assurance: A Resilience Taxonomy*, September 2015, p. 3. The white paper's definition of *resilience* was adopted from DoD Directive 3100.10, *Space Policy*, Office of the Under Secretary of Defense for Policy, August 30, 2022.

[8] Office of the Assistant Secretary of Defense for Homeland Defense and Global Security, 2015, pp. 6–8.

- **Disaggregation**: "the separation of dissimilar capabilities into separate platforms or payloads"[9]
- **Distribution**: "utilizing a number of nodes, working together, to perform the same mission or functions as a single node"[10]
- **Proliferation**: "deploying larger numbers of the same platforms, payloads or systems of the same types to perform the same mission"[11]
- **Diversification**: "contributing to the same mission in multiple ways, using different platforms, different orbits, or systems and capabilities of commercial, civil, or international partners"
- **Protection**: "active and passive measures to ensure those U.S. space systems, and those of our partners upon which we rely, provide the required quantity and quality of mission support in any operating environment or condition"
- **Deception**: "measures taken to confuse or mislead an adversary with respect to the location, capability, operational status, mission type, and/or robustness of a national security system or payload."

While it provides a definition of resilience and the ways in which it could be achieved, the 2015 white paper does not provide guidance on how to measure resilience. For our research, we adopted the metrics shown in Table 1.1. These are categorized as measures of how well the system responds to threats, where *system* includes both equipment (space- and ground-based) and personnel. As noted in the last column of the table, these measures of resilience must be informed by operational context (e.g., percent of capability needed, operationally relevant recovery time) if analysts are to create well-defined metrics—hence this research's emphasis on

---

[9] An individual satellite is a platform. The equipment that provides services is termed the *payload*, denoting a function a customer is willing to pay for. An individual satellite can host multiple payloads, but only if those payloads have compatible requirements for power, thermal control, orbital control, and, if disaggregated, military capability.

[10] The distribution example given in the white paper is that of the global positioning system (GPS), which uses multiple satellites in different planes and a series of monitoring stations around the globe to provide worldwide location and timing services. If any single GPS satellite or monitoring station is lost, the global services are still available, though perhaps with slightly degraded accuracy.

[11] Distribution and proliferation are quite similar. The primary distinction is that in a distributed architecture, the platforms, payloads, and ground elements work together as nodes in a network to assure the quality of the overall service; in a proliferated architecture, the quality of service is only dependent on the overhead satellite and does not depend on those that are out of view.

When continuous service is a desired quality, as it is for communication services, a proliferated low earth orbit communications satellite network must be interconnected in a mesh, making it both proliferated and distributed. The mesh can be provided using crosslinks in orbit or interconnects on the ground. A crosslinked architecture is resilient to ground-based attacks, and a mesh using ground-based interconnects is resilient to in-orbit attacks. Both are vulnerable to downlink jamming attacks, and their large attack surface makes them vulnerable to cyber intrusion. This example illustrates that resilience is as much a function of the attack vector as it is of the architecture of the system.

developing a repeatable methodology that can define, assess, and evaluate resilience *as operational context changes*.

Table 1.1. Metrics of Operational Resilience

Category	Definition	Unit of Measure	Issues
Avoidance	The probability of deterring or otherwise avoiding the threat or its impact	Direction of change in probability	Difficult to measure, highly subjective
Robustness	The degree to which the impact of an attack can be absorbed	Percent of capability lost if threat manifests	Requires defining what *100 percent capability* means
Recovery, short term	Post attack—how much capability can be recovered in an operationally relevant time frame	Percent of capability recovered in time	Requires defining an operationally relevant time frame
Recovery, long term	Post attack—how long it takes to recover to a robust posture	Time	Requires defining what a *robust posture* means

SOURCE: Metrics adapted from Ron Burch, *Resilient Space Systems Design: An Introduction*, CRC Press, 2019.

Assessing Resilience

For our research, we focused on operational resilience as opposed to architectural resilience. We are interested in factors regarding partnerships that manifest at the time of battle and impact how commanders use the combined space architectures that the partners bring. These decisions can impact the strategic, operational, and tactical levels of military planning and execution. In general, operational resilience arises when commanders use diversification (of doctrine, plans, and tactics), protection, and deception.

Earlier work at the RAND Corporation focused on how the United States could use non-materiel methods (i.e., methods other than adding new satellites, payloads, and systems) to improve space operational resilience. In that work, the researchers created a network diagram of the interactions among doctrine, organization, training, materiel, leadership and education, personnel, facilities, and policy (DOTmLPF-P) to study how different investments in any of those areas would impact a space system's operational resilience.[12] The researchers held a series of meetings with SMEs to rate the effectiveness of a particular investment in countering a given threat. A snippet of a diagram illustrating those ratings is shown in Figure 1.1, where purple and blue denote the most effective counters; green is effective; yellow is marginal; and red and white indicate that the investment has little to no impact on mitigating the threat.

---

[12] See Gary McLeod, George Nacouzi, Paul Dreyer, Mel Eisman, Myron Hura, Krista Langeland, David Manheim, and Geoffrey Torrington, *Enhancing Space Resilience Through Non-Materiel Means*, RAND Corporation, RR-1067-AF, 2016. See also Paul Dreyer, Krista Langeland, David Manheim, Gary McLeod, and George Nacouzi, *RAPAPORT (Resilience Assessment Process and Portfolio Option Reporting Tool): Background and Method*, RAND Corporation, RR-1169-AF, 2016.

Figure 1.1. Example of Ratings Provided by SMEs for Use in Resilience Analyses

Options / Threats	Baseline	Develop tactics for likely counterspace threats in advance of their deployment	Develop a more timely anomaly resolution process
ASAT attack			
Jamming uplink			
Blinding laser/dazzling attack			

Color Key:	None (0)	Low (0.1)	Low/Medium (0.25)	Medium (0.5)	Medium/High (0.75)	High (1)
Baseline Meaning	No capability on threat impact	Little capability on threat impact	Minimal capability on impact	Noticeable capability on threat impact	High added capability on threat impact	Very significant capability on threat impact
Option Meaning	No added capability on threat impact	Little added capability on threat impact	Minimal added capability on impact	Noticeable added capability on threat impact	High added capability on threat impact	Very significant added capability on threat impact

SOURCE: Adapted from Dreyer et al., 2016.
NOTE: ASAT = anti-satellite.

These ratings represent a snapshot of how experts think about a problem at a given point in time, but we hypothesize that there are several undocumented assumptions in the rating that might change with time and circumstance. Therefore, our research goal is to develop a method that will help segregate factors that change with

- community attitudes and doctrine
- the operational context of the conflict
- assumptions about the effectiveness of specific training or employment of specific tactics, techniques, and procedures (TTP) in mitigating the effect.

If we are successful at segregating those factors, our next goal is to validate the underlying assumptions. There are few methods that attempt to validate community attitudes or how those attitudes change with time versus those attitudes that change because the operational context is different. Methods to establish the effectiveness of TTP via training and exercise are relatively well developed and are ingrained in U.S. military readiness reporting. Therefore, for the research reported here, we focused on developing a methodology to further characterize the first two factors that change with time: community attitudes and operational context. In our exemplar network analysis, we also focused on providing insights as to the sensitivity of subsequent resilience analyses that use those factors.

Research Questions

The above research goals led us to formulate the following research questions:

1. What factors influence the currency of information provided by SMEs for operational resilience analyses?
2. Can we develop a method to distinguish factors that change with community attitudes or the operational context assumed by SMEs when they are rating resilience impacts (or both)?

Organization of This Report

Chapter 2 documents the methodology we developed by which resilience criteria can be defined, assessed, applied to decisions, and evaluated *over time,* with a particular focus on the qualitative assessments of SMEs. Chapter 3 then details the results of applying that methodology to the question of integrating coalition and commercial partners into MILSATCOM missions. Chapter 4 summarizes the findings and recommendations of the research.

Appendix A contains a discussion of different partnership constructs that have been used or might be used in MILSATCOM missions. Appendix B then documents an exemplar quantitative network-based resilience analysis comparing the expected impact of two of those partnership constructs: a fully integrated partnership and a hosted payload partnership. Appendix C contains the interview protocol we used to solicit possible logic models for how integrating partners might improve USSF resilience. Finally, Appendix D documents the survey we used to measure community attitudes.

Chapter 2. The Methodology

Overview

The methodology that we developed by which resilience criteria can be defined, assessed, applied to decisions, and evaluated *over time,* with a particular focus on the qualitative assessments of SMEs, is illustrated in Figure 2.1. The method begins with a series of semistructured interviews designed to elicit mental models about how and why integrating allies and partners into MILSATCOM missions might lead to greater resilience. Specifically, the intent of the interviews is to elicit an expression of a theory of change that links inputs to outcomes for a given initiative.[13] For example, we hoped to elicit such statements as "integrating allies into the mission provides the coalition with additional options X that improve this Y aspect of resilience." Attempting to elicit mental models in a way that does not prejudice the interviewee is difficult: The interview protocol must accommodate both interviewees (1) who have a well-developed theory of change, such as those working in the Office of the Secretary of Defense for Space Policy, and (2) who have a more intuitive theory of change based on their operational and tactical experience.

The second step of the methodology is to take the factors gleaned from the interviews and construct a survey instrument to measure whether the broader MILSATCOM community believes hypothetical factors X do in fact lead to Y. For this, we use a technique known as *factor analysis* to ascertain the independence of factors related to community attitudes about a given product or service. In essence, a factor analysis tells us whether the factors we have hypothesized as impacting community attitudes are, in fact, independent variables or whether some other latent factor is the independent variable. Some statistical packages use complex equations based on eigenvalues to determine independence of variables, but our methodology uses a simple graphical analysis.[14]

It is not enough, however, to simply identify the independent variables that influence whether and how integrating allies and partners into the MILSATCOM mission impacts resilience: The methodology must also measure the shift in community attitudes about those factors that arise due to the operational context. Therefore, our survey design incorporates two vignettes that we asked respondents to consider when answering the survey questions. While all respondents were

---

[13] These mental models are often referred to in the literature as *logic models*, *roadmaps*, or *theory of change*. An excellent overview is provided in Center for Community Health and Development, *Community Tool Box, Chapter 2, Section 1: Developing a Logic Model or Theory of Change, Main Section*, University of Kansas, undated.

[14] For an overview of these statistical methods, see Stephanie Glen, "Factor Analysis: Easy Definition," webpage, Statistics How To, undated. See also Timothy A. Brown, *Confirmatory Factor Analysis for Applied Research*, 2nd ed., Guilford Press, 2015.

asked to answer within the context of a fully integrated partnership arrangement, half of our respondents were asked to answer a subset of the questions within the context of a more highly constrained hosted payload construct, where the partnership is more circumscribed by contractual and operational barriers.

The last step in our methodology is to weight the independent factors (the inputs) derived from the community attitudes survey and use those factors in a quantitative analysis to produce insight on the type of resiliency produced and sensitivity to these more ephemeral community attitudes–based factors. An exemplar of such analysis is documented in Appendix B, but given the strength of the confounding factors found in the factors analysis, we do not draw conclusions from that analysis.

Figure 2.1. A Methodology for Defining, Assessing, Applying, and Evaluating Resilience Criteria as Community Attitudes and Operational Context Changes

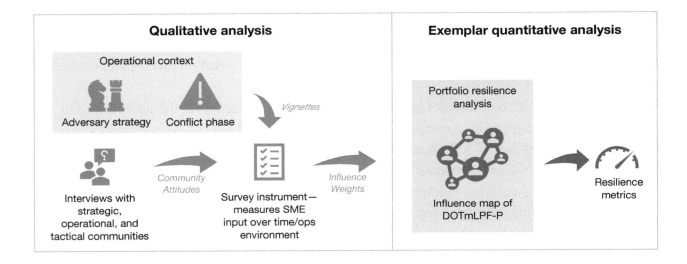

Eliciting Logic Models

Logic Models and Why They Matter

As discussed earlier, logic models express a theory of change. They are useful not just for helping a community explicitly articulate how change will occur but also for guiding newcomers in understanding how a particular change effort or initiative is intended to operate and why. A commonly articulated model that we found in our work regarding how or why integrating partners into the MILSATCOM mission might lead to more resilient operation is illustrated in Figure 2.2. According to this logic model, partners bound by diverse doctrines provide a greater freedom of action across the coalition. What one partner cannot do, another might be able to do. In turn, this greater freedom of action is thought to aid resilience. Two caveats must be noted:

- The logic model is incomplete because it does not delineate whether the greater resilience is because of the coalition's ability to deter (avoid) attacks, absorb attacks, or recover from attacks. Instead, it implies that diversity of doctrine might improve all aspects of resilience.
- The logic model also says nothing about the value of closely aligned doctrines. It is equally possible that doctrinal alignment could allow a coalition to present a unified front or anticipate each other's responses in ways that improve resilience.

Figure 2.2. Example of a Logic Model

Logic Model: "Different but synergistic doctrines can be leveraged to provide greater freedom of action and thus more resilient operations."

In general, logic models change over time. This is because they are hypotheses of how change will occur, not proven models. It is also because every strength is likewise a weakness that an adversary will find a way to exploit. In our example in Figure 2.2, if an adversary can exploit doctrinal diversity between partners to the adversary's advantage (i.e., exploit the seams in the coalition), community sentiment regarding the benefits of diversity in doctrine might radically change. As new experiences change the community's assessment of what does and does not work in the real world, logic models inevitably are updated. Although logic models do not always evolve in the direction of truth, a shift in logic models might be a reliable indicator that something about the operational context has changed. Given that our goal is to provide the USSF with an indicator that change has occurred and, therefore, that prior analyses that relied on those logic models must be updated, the ability to reliably and repeatably elicit logic models that can in turn be analyzed for change is a key component of our methodology.

Interview Design—Eliciting Logic Models

As discussed above, logic models are simply theories of how change might occur. We anticipate that a theory expressed by an SME is based on the sum of their life experiences. To provide generalizable knowledge, a variety of experts, each with their own experience, must be consulted. However, even then, the outcome of an elicitation of logic models must be seen as either an expression of *today's* conventional wisdom or as a means to understand the range of theories of change that could impact *future* outcomes. In our research design, we are interested in eliciting logic models that provide the range of theories of change that could impact future outcomes. Only later, during the survey step, did we begin to measure conventional wisdom and, even then, only within the context of discovering which factors are independent inputs and whether we can measure *change* in the conventional wisdom.

To that end, our interviews were structured using open-ended questions that allowed our interviewees to articulate their theories with as little guidance from the interview team as possible. We also pre-planned a set of questions that gradually narrowed the context of the interviewees to the specific questions of how integrating coalition and commercial partners into the MILSATCOM mission might impact the resulting resilience of coalition operations. We used these more specific questions only if an interviewee had difficulty narrowing their context in a way that allowed them to articulate their theory of change. We also designed our interview protocol to allow and encourage participants to think about the issue at the strategic, operational, and tactical levels of warfare. The interview protocol can be found in Appendix A.

Given our goal of obtaining the widest variety of logic models, our list of candidate interviewees emphasized diversity over quantity.[15] We wanted insights from across the DoD MILSATCOM community at different levels of warfare (strategic, operational, and tactical), from close U.S.-allied nations (Australia and Canada), and from the commercial sector that supplements DoD's military communications capabilities. To ensure we obtained the views of not just current but also possible future commercial partners, we wanted input from both current and potential DoD SATCOM providers. Table 2.1 provides the number of interviews we conducted as a function of partner type (DoD, allied, commercial) and the interviewees' expertise in levels of warfare (strategic, operational, tactical). Interviewees who had expertise at multiple levels of warfare were counted multiple times.

Table 2.1. Interview Participant Coverage Across the Levels of Warfare

	DoD	Allied	Commercial
Strategic	4	2	1
Operational	6	2	2
Tactical	3	0	0

Interview Response Analysis

In analyzing the interview responses, we tasked two independent reviewers to ensure personal biases did not unduly influence our results. These reviewers focused on identifying themes, which we call *dimensions* of resilience. For each dimension, we then developed a pair of possible logic models to cover the range of future experience. For example, for the doctrine dimension, despite our interviewees' focus on the value of *diversity* in doctrine, we also developed a logic model for how *aligned* doctrine can contribute to resilience. A summation of

---

[15] Coverage, not quantity, is the important metric when interviewing to establish the variety of thought on a given topic.

the dimensions, mapped against the level of warfare and impacted resilience metric, is provided in Table 2.2. In this table, we group the dimensions by the basic questions applicable to any contemplated partnership:

- Should DoD integrate in coalition and commercial partners?
- If yes, what capabilities should be integrated?
- If yes, what enables successful integration?

Table 2.2. Dimensions of Resilience Developed from Our Analysis of the Interviews

Basic Question	Dimension: Range of Possible Community Attitude	Level of Concern	Operational Resilience Metric Impacted
Should DoD integrate allies and partners?	Dependencies: constraint or opportunity	Strategic	Avoidance
	Interoperable situational awareness (SA): robustness or shared vulnerability	Strategic	Robustness
	Planning and resource allocation: timeliness or complexity	Operational	Short-term recovery vs. Robustness
What should be integrated?	Integration: augmentation or core capability	Strategic	Short-term recovery
	Integration: tailored vs. full spectrum capability	Operational	Avoidance, robustness
	Integration: diversity vs. proliferation	Tactical	Short-term recovery
What are the enablers of integration?	Doctrine: complimentary or aligned	Strategic	Avoidance, robustness
	Organizations: autonomous vs. integrated operations	Operational	Robustness, short- and long-term recovery
	Training: just in time or sustained	Tactical	Short- and long-term recovery

Paired logic models that might explain the range of possible community attitudes regarding those dimensions are provided in Table 2.3.

Table 2.3. Dimensions of Resilience and Explanatory Paired Logic Models

Dimension: Range of Possible Community Attitudes	Paired Logic Models
Dependency	• Fighting in a coalition creates opportunities to confound adversary decisionmaking. • Fighting in a coalition creates dependencies on partners that constrain national decisionmaking.
Interoperable SA	• Shared situational awareness from multiple sources ensures that commanders have the information they need to make decisions. • Local situational awareness that does not rely on external data ensures that commanders have reliable information, allowing rapid decisionmaking.
Planning and resource allocation	• Partnerships that have clear boundaries but separate planning can respond quickly, leading to more resilient operations. • Partnerships that are fully meshed with integrated planning provide a clear understanding of total system capability, leading to more resilient operations.
Integration: augmentation or core capability	• To assure resilient space operations, the United States should supply all of the core capabilities needed for operations. • To assure resilient space operations, the United States should augment core capabilities with those from partners.
Integration: tailored vs. full-spectrum capability	• Operational resilience can best be achieved if partners bring fully capable systems that are interchangeable with U.S. systems. • Operational resilience is best achieved if partners bring less capable but interoperable systems.
Integration: diversity vs. proliferation	• Minimizing the diversity of resources simplifies the coalition's ability to reallocate resources and achieve resilience in operationally relevant time frames. • Adding greater diversity of resources complicates the adversary's decisionmaking, resulting in a more resilient architecture.
Doctrine as an enabler	• Having similar doctrines leads to stronger partnerships and more resilient operations. • Different but synergistic doctrines can be leveraged to provide greater freedom of action and more resilient operations.
Organization as an enabler	• A more loosely integrated command and control system is more operationally responsive. • Fully integrated command and control systems allow planners to use all systems to best effect, resulting in a more robust architecture.
Training as an enabler	• Repeated joint exercises with partners over many years are the best way to ensure effective partnerships in the field.[a] • Immediate pre- or post-engagement joint exercises with partners are the best way to ensure effective partnerships in the field.

[a] Our use of the word *joint* reflects the common definition of "involving the united activity of two or more" rather than the more specific definition of "constituting an activity, operation, or organization in which elements of more than one armed service participate" (Merriam-Webster, "joint," dictionary entry, undated).

Measuring Community Attitudes

Sample Size

The goal of the survey design used in this methodology is not to obtain measures of consensus on topics of resilience but only to ascertain if hypothesized factors impacting resilience are independent. This latter goal requires a significantly smaller sample size than the former. As with our interviews, we sought respondents from across the range of strategic, operational, and tactical communities and from allies and commercial entities. Respondents had strict anonymity, and we collected only the minimum demographic data needed to ensure that no one experience base skewed our results. In total, we had 22 respondents.

Survey Design to Measure Factor Independence

In the design of the survey, the paired logic models documented in Table 2.3 are used to ascertain whether the paired items are independent factors that should be modeled as separate inputs in resilience modelling. Each of the logic models in the pair is formulated as a positive statement.[16] Survey respondents are then asked to rate each statement on a Likert scale of *strongly agree, agree, neutral, disagree, strongly disagree*.

For example, the paired positive statements of the logic models regarding the impact of doctrinal diversity and doctrinal alignment are:

- Having similar doctrines leads to stronger partnerships and more resilient operations.
- Different but synergistic doctrines can be leveraged to provide greater freedom of action and more resilient operations.

These are not necessarily mutually exclusive statements: An intelligent and thoughtful respondent might have the same sentiment regarding each of these statements, opposing (mirrored) sentiments, or totally independent sentiments about them. It is the lack of correlation between sentiments about these statements (whether that correlation results in a similar or mirrored response distribution), that indicates the logic models express independent factors. These paired statements are deliberately placed on the same screen when viewed by our survey respondents. This proximity predisposes respondents to think of the statements as dependent pairs rather than independent factors. A lack of correlation in the responses is thus a stronger indicator of independence than might be obtained were these statements on separate screens.

To accommodate those cases where later analysis of survey results indicate that the paired factors are not, in all likelihood, independent, we also hypothesized factors that might be the

---

[16] Positive statements should be used to express both logic models to avoid cognitive biases associated with negative statements. Social scientists have long known that a negative statement will invoke stronger sentiments from survey respondents. In addition, neutral language should be used, and words that might trigger individual biases should be avoided. We asked three different reviewers with multiple years of experience in the MILSATCOM community to review our language for neutrality.

actual independent factor. We term these candidate alternate independent factors as *orthogonal* to the dimension of resilience the paired logic models expresses.[17] For the orthogonal factor, we again make one or more statements that embody the hypothesized alternative logic model.[18] For our example of doctrinal impacts, we hypothesized that perhaps it is operational (not doctrinal) diversity that is the independent factor. To that end, we included a third statement in the survey:[19]

- Diverse thought on how to conduct operations is critical to achieving resilience.

As before, survey respondents are asked to rate this statement on a five-point Likert scale from *strongly agree* to *strongly disagree.*

It is critical to note that our analysis does not attempt to measure consensus of sentiment. Although there are a few instances in our survey results where the consensus is so overwhelming as to be statistically significant, measuring that consensus is not the intent of the survey. Instead, the intent of the survey's design is to detect a lack of correlation in the distribution of responses. It is the lack of correlation that determines which factors might be independent and should be included in a resilience analysis.

Survey Design to Understand Trust as a Confounding Factor

There are two additional factors that our research leads us to believe have the potential to invalidate (i.e., confound) the factors analysis described above.[20] The first is trust, which is an undercurrent that ran throughout the interview responses. On detailed analysis of the interviews, we decomposed trust into two elements:

- Trust in the overall DoD system to provide sufficient resources needed to conduct resilient operations in space

---

[17] In math, to be orthogonal is to be at right angles to a reference or to be statistically independent. Our use of the term in developing candidate alternative independent factors is not to be interpreted as a mathematical orthogonality but rather a logical orthogonality: thinking about a dimension of resilience from an alternate, but not opposing, point of view. These candidate alternatives were expressed in the interviews but did not necessarily rise to the level of a consistent theme.

[18] In some cases, the hypothesized orthogonality was another logic pair.

[19] Single orthogonalities were grouped with the logic model pair so that lack of correlation provides a stronger signal of independence.

[20] A confounding factor is one that is not considered during hypothesis (cause and effect) formation but that fully influences both cause and effect in a way that invalidates the hypothesis. A simple example of a confounding factor is as follows: Suppose you are exploring the relationship between shark attacks and ice cream sales and find that shark attacks do indeed precede an uptick in ice cream sales. The confounding variable here is temperature: There are more sharks in warm water and more ice cream sales in warm weather. A good layperson's explanation of confounding factors can be found in Zach Bobbitt, "What is a Confounding Variable? (Definition and Example)," webpage, Statology, February 19, 2021.

Our orthogonalities are also examples of confounding factors, but they only impact one dimension of resilience. In this report, we reserve the term *confounding factor* to those factors that influence the factors analysis across the majority of the dimensions of resilience denoted in Table 2.2.

Resources, in this case, could include spacecraft and the gateways to access those spacecraft, but more often, interviewees referenced the planning tools, tactics, procedures, and training needed to fully utilize those spacecraft and gateways. This lack of trust could confound the factors analysis in two ways:

- If our respondents do not trust that the United States will have sufficient spacecraft and gateways needed to conduct resilient operations, then integrating coalition and commercial partners is likely to be seen as a benefit, independent of the more-nuanced logic models inherent in our factors analysis.
- If our respondents do not trust that they will have the tools and training to effectively plan and utilize capabilities, then integrating coaltion and commercial partners into space operations is unlikely to be seen as a benefit, independent of what the partnership brings to the warfight.

- Trust in the partnership to be able to execute resilient operations

We believe it might be important to understand the directionality of this trust. If trust is not symmetric, then we might expect to see very different sentiments regarding factors analysis as a function of whether DoD or the partner is responding.

Therefore, the survey includes a set of questions to probe how our respondents think about these trust issues. Again, these are positive statements that respondents rate on the five-point Likert scale from *strongly agree* to *strongly disagree*.

This first group of statements were only rated by DoD personnel and focus on trust in the DoD system to provide the resources necessary for resilient space operations:

1. The United States has enough resources to achieve robust operations, independent of partners.
2. The United States has the diversity of resources needed to achieve robust operations, independent of partners.
3. The United States has the tools and procedures needed to reallocate resources to achieve resilience in operationally relevant time frames.
4. The United States has the tools and procedures needed to incorporate diverse resources into planning, thus enabling a more robust architecture.

This second group of statements were rated by DoD and allied military personnel to test the symmetry of trust within the coalition:

1. Coalition partners can rely on the United States for rapid response to overcome adversary threats.
2. The United States can rely on coalition partners for rapid response to overcome adversary threats.

Finally, the third group of statements were rated by all respondents and are designed to provide a more nuanced understanding of the nature of trust between the military and commercial partners. In this case, the trust is not anticipated to be symmetric because the partnership is not symmetric. Our statements presuppose that the partnership is a provision of

SATCOM services by the commercial entity with the expectation of defense provided by the military entity.

1. The military can trust commercial partners to strive to continue to provide service while overcoming adversary threats.
2. The military can trust commercial partners to provide robust services that can withstand attack.
3. Commercial partners can trust the military to defend them when they come under attack.

Survey Design to Understand Beliefs as a Confounding Factor

Beliefs are a second potential confounding factor. As we noted earlier, although the U.S. military has a taxonomy of relience, it does not have a standard measure of resilience. From our research, we defined four measures relative to an attack: avoidance (the ability to deter an attack), robustness (the ability to absorb an attack), short-term recovery (post-attack), and long-term recovery (post-attack), where the discriminator between short and long term is the operationally relevant time frame. If our respondents have strong beliefs about the importance of each of these measures to overall resilence, then it might impact their sentiments regarding the statements we use for our factors analysis. Some factors impact some measures much more strongly than others, and it could be that our respondent's beliefs about the importance of the factors is coloring their overall responses. To that end, the survey asks our respondents to rank order a set of statements about how resilience is "best" achieved:

> For each group of statements below, rank order them to indicate how strongly you agree. Use a 1 to indicate the one you most agree with, a 2 to indicate the statement you agree with next, etc.
>
> 1. Resilience for the SATCOM mission is best achieved through the ability to:
>
> _____ recover quickly after an adversarial action impacts service
>
> _____ avoid any service impact from adversarial action
>
> _____ minimize the service impact of adversarial action
>
> _____ finely tune the system in response to adversarial action

There are two other factors from our interviews that we believe might be confounding and that are best suited to a rank-order questionaire: (1) beliefs about the importance of possible outcomes of joint training and (2) beliefs about the factors that lead to a successful partnership.[21] To that end, survey respondents are asked to rank order the following two statements:

> 1. The most important outcome of joint exercises with partners is:
>
> _____ the informal bonds formed by training together.

---

[21] These beliefs would not confound all factors analysis, but they do have the potential to confound more than one of our dimensions, and the range of possible logic models was more extensive than could be expressed with simple paired statements.

_____ a joint understanding of each party's tactics, techniques, and procedures.

_____ the expertise gained by training together.

2. The success of a partnership is most dependent on both parties:

_____ being transparent about their goals, objectives, capabilities, and constraints

_____ believing the other has their interests at heart.

_____ having the capability to effectively act in the joint interest.

When rank-order questions are used, it is vitally important that researchers not imply an absolute valuation to the response—this is merely a relative order. Because a rank order does not provide any information about the strength of sentiment regarding these issues, these responses are not suitable for any type of factor analysis. However, it does force respondents to make choices (i.e., a rank choice precludes ties). These choices are value judgements about what is _best_ or _most_ important or determinate.

Survey Design to Detect Change in Community Attitudes

The final element that influenced our survey design is the desire to determine the survey's usefulness in detecting changes in community attitudes that arise from changes in the operational context. As we noted earlier, the goal of the methodology is to provide the USSF with an indicator that change has occurred and, therefore, that prior analyses that relied on logic models or factors that influence the logic models must be updated. If survey responses do not change when operational context changes, the survey is a failed instrument. We hypothesized that operational context is provided by the phase of war, the adversary tactics, and the constraints a particular partnership construct might impose on the warfight. To that end, half of the survey respondents are asked to consider an alternate vignette when providing Likert ratings to some of the factors analysis statements. The two vignettes used in the survey instrument are shown in Table 2.4.

Table 2.4. Vignettes Providing Operational Context

Vignette Use	Vignette
All respondents	U.S. service members and partners are collocated in theater, fighting in a coalition command structure. While different partners have brought their own SATCOM terminals and gateways, any nation's communication systems (including communication services procured from commercial providers) can be used to support the coalition within pre-negotiated prioritized limits. The coalition partners have trained together to respond to contingencies and work together in a joint SATCOM planning center to redistribute communications between their various systems in response to adversarial actions or rapid changes in the operational environment. Currently the adversary is jamming several of the coalition's pre-planned communication channels. Multiple systems are available to fill the need for communications services, but replanning is necessary to establish new communication links that would circumvent the jamming attack
Additional for 50% of respondents	A military payload has been hosted on a commercial satellite constellation to provide robust coverage at minimal cost. The commercial provider manages the satellite constellation and operates several (nonmilitary) payloads. The USSF operates the military payload using an in-band communications link. If the in-band communication link to the military payload is disrupted or degraded, the commercial operator can utilize alternate communication links—on request and on a noninterference basis—to assist the military operations center with debug. For security reasons, connectivity between the military and commercial operations centers is limited to email, phone and secure file transfer

The additional vignette was deliberately constructed to detail realistic operational constraints that arise when a military payload is hosted on a commercial satellite. Because constraints are generally viewed as having a negative impact on resilience, we expect that respondents who have been provided this additional information might view partnerships less favorably: It is this shift we seek to measure. We provided the additional vignette toward the end of the survey, before our questions about partnership issues but after a baseline has been established. This allows us to examine how the vignette changes collective sentiments between groups and how it might shift the sentiments of an individual respondent. To minimize survey fatigue, we were careful to reshape statements about any repeated dimension or confounding factor to keep them fresh. In doing so, there is a small risk that we might change the statement in a way that provokes a different sentiment. Therefore, small shifts in an individual's rankings of a dimension before and after viewing the additional vignette should not be construed as having been induced by the more constrained vignette. Only large and distinct shifts are considered significant.

Survey Results Analysis—Factor Relevancy and Sensitivity to Operational Context

Factors analysis is conducted as follows:

1. The response distributions of the paired set are examined to determine if there are significant differences in the pattern of distribution. If there are, then the logic models express independent variables that should be included in later resilience analyses.
2. For questions in which half the respondents were given the additional vignette information, the response distributions segregated by vignette group are examined. If the vignette appears to have significantly altered the distribution of responses, we can be

18

more confident that the survey instrument can serve as a valid indicator of change in community attitudes.

3. For selected questions in which we suspect that the respondent's background might impact the distribution of responses, the response distributions of military versus commercial versus DoD versus partner are examined.

4. If the distribution of the responses is not clearly distinct, the final step is to

   a. check the orthogonality to determine if it is the confounding variable
   b. check for trust and other confounding beliefs.

Although there are statistical packages for checking whether two distributions of responses are correlated, we chose to use a graphical analysis. Examples from the actual data set obtained from our survey are shown below.

The first example, shown in Figure 2.3, is for a pair of logic statements that yields a very different response distribution pattern. The statements are

1. Partnerships that have clear boundaries, but separate planning, can respond quickly leading to more resilient operations.

2. Partnerships that are fully meshed with integrated planning provide a clear understanding of total system capability leading to more resilient operations.

Although the responses to our statement that clear boundaries but separate planning improves resilience is fairly uniform, the response to our statement on the benefit of fully integrated planning has a distinctly positive skew. Because of the lack of correlation between the distributions, we conclude that the factors of clear boundaries and degree of integration in planning appear to be independent.

Figure 2.3. Uncorrelated Response Distribution Example

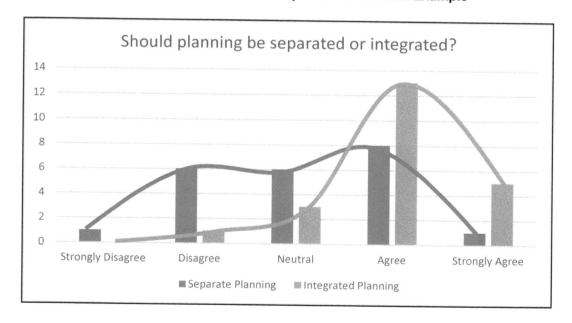

When we look at these paired logic statements by vignette (see Figure 2.4), our insights become more nuanced and confirm that these factors are independent. Those respondents who viewed the more constrained hosted payload vignette (Vignette B) respond to the statement on separate planning with a clearly bimodal distribution: Most respondents either agree or disagree, with few being neutral. Meanwhile, respondents who only viewed the more general partnership vignette (Vignette A) mostly disagree or are neutral on the subject of separate planning. All respondents, regardless of vignette viewed, have a positive sentiment regarding our statement on the benefit of integrated planning.

Figure 2.4. Response Distribution by Vignette Example

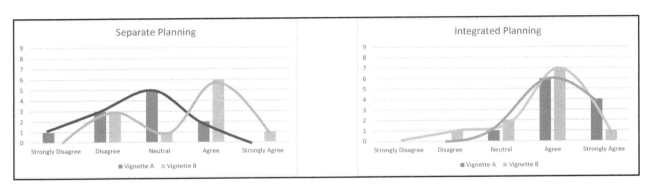

Figure 2.5 shows a more ambiguous case that is the result of our statements about the value of diverse but synergistic versus aligned doctrine. Recall that the statements made are

1. Having similar doctrines leads to stronger partnerships and more resilient operations.
2. Different but synergistic doctrines can be leveraged to provide greater freedom of action and more resilient operations.

The distributions of the responses to these statements are not substantially different: Both exhibit a slight bell curve and, though they have different means, could imaginably be samples from the same distribution. Therefore, to examine the distributions further, Figure 2.6 plots the responses by military respondents (who presumably might have a more nuanced view of the benefits of doctrine, whether similar or different) and by commercial entity respondents.

Figure 2.5. Ambiguous Response Distribution Example

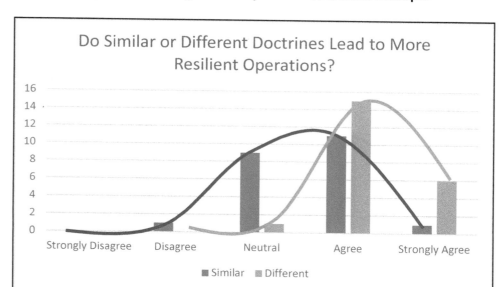

Figure 2.6. Response Distribution by Military vs. Commercial Example

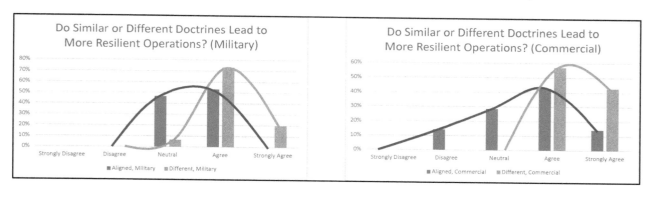

In Figure 2.6, it becomes more apparent that, at least for commercial respondents, these logic models have different distributions. Therefore, in our later quantitative analysis, we consider freedom of action as a different (and positive) factor from doctrinal partnership ties.

In some cases, we find that our respondents view our paired logic models as being a highly correlated either/or construct. In these cases, the logic models are clearly not independent but are two sides of the same coin. The clearest such case is for the statements we made concerning organizational constructs for command and control (C2): Response distributions are shown in Figure 2.7.

1. A more loosely integrated command and control system is more operationally responsive.
2. Fully integrated command and control systems allow planners to use all systems to best effect, resulting in a more robust architecture.

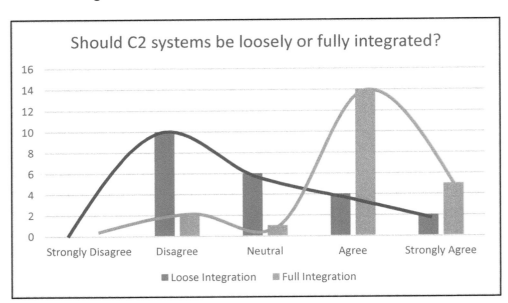

Figure 2.7. Correlated and Mirrored Sentiment Example

In constructing these statements, we had hypothesized that respondents might value both the operational responsivity provided by loose integration and the robustness offered by full integration. Instead, respondents overwhelmingly rejected our statement regarding the benefits of loose integration. To determine whether this rejection was because they value robustness over responsivity, we check the statements we had made about which measures of resilience are most important. The scored responses to our ranked questions are shown in Table 2.5.

Table 2.5. Responses to Ranking of Resilience Measures

Resilience Measure	Number of Top Choice Votes	Number of 2nd Choice Votes	Number of 3rd Choice Votes
Recover quickly after adversarial action	4	9	7
Avoid service impact of adversarial action	12	2	5
Minimize service impact of adversarial action	6	11	6
Finely tune the system in response to adversarial action	1	1	5

The data in this table are consistent with the idea that robustness, the ability to avoid or minimize service impacts of an attack (combined 18 first place votes), is valued more highly than responsiveness, the ability to recover quickly or to fine tune a response to an attack (only five first place votes). The confounding variable of beliefs about the value of different measures of resilience might have greatly influenced the response to these paired logic models about organizational ties. Therefore, in later analysis, we chose to model organizational ties as a random variable that could be either positive or negative.

There is one other logic model pair that might have been confounded by this preference for robustness over responsiveness—our statements about diversity of resources:

1. Minimizing the diversity of resources simplifies the coalition's ability to reallocate resources and achieve resilience in operationally relevant time frames.
2. Adding greater diversity of resources complicates the adversary's decisionmaking resulting in a more resilient architecture.

Indeed, an examination of the response distributions (shown in Appendix C) shows mirrored and perhaps correlated sentiments. Our statements about greater diversity that complicates the adversary's decisionmaking (and thus presumably improves robustness) received strong agreement, while our statement about minimizing diversity to improve responsivity received strong disagreement.

Before proceeding to a discussion in Chapter 3 on how to use the insights gained from analysis of the survey responses, there is one last response distribution pattern to discuss. When respondents are asked to respond on a Likert scale to a statement they regard as being not applicable but are not given a "not applicable" response option, they will instead respond with the "neutral" option. Therefore, best practice is to either include a "not applicable" option or, if that was not done (as it was not in our case), flag any distribution that looks like that shown in Figure 2.8 and discard these inputs in further analysis.

Figure 2.8. "Not Applicable" Response Distribution Example

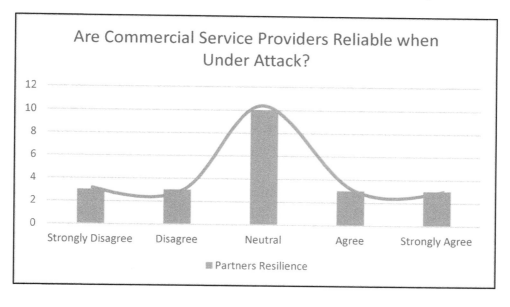

We encountered this pattern only once in our dataset, in response to our statement that "[t]he military can trust commercial partners to strive to continue to provide service while overcoming adversary threats." It seems clear that most respondents did not feel that they had sufficient context to have strong sentiments about this question.

Summary

In this chapter, we described a methodology, beginning with semistructured interviews designed to elicit logic models, or theories of change, regarding how integrating U.S. MILSATCOM missions with coalition and commercial partners might impact resilience. We discussed how we selected our interview participants with an emphasis on diversity of thought and how we structured the interview questions to elicit a *range* of possible ways that partnerships might impact resilience while using neutral language to avoid inserting our own biases. The interview protocol is provided in Appendix C.

The next step in the methodology is to analyze the interviews for themes. This resulted in the identification of nine dimensions of resilience that impact three basic questions about partnerships: (1) Should the United States partner? (2) What capabilities should be integrated in the partnership? and (3) What enables a successful partnership? For each of the nine dimensions, we identified the metric of operational resilience that the dimension would impact and developed sets of paired logic models to explore community attitudes regarding that dimension. Two important aspects of the logic model construction are that (1) we stated all outcomes in a positive way to avoid unconscious bias, and (2) although the logic pairs explore opposing ends of the spectrum regarding a particular dimension, the statements are not necessarily mutually exclusive. To assess whether underlying factors are independent, it must be possible for a thoughtful respondent to agree (or disagree) with both statements.

We then described how we used the paired logic model statements as the basis for a community attitude survey to better understand the degree to which SMEs agree or disagree with logic models and how those attitudes might change with circumstance. The survey was also designed to explore attitudes about two possible confounding factors that had emerged from the interviews. The first was trust in both the partnership and the U.S. ability to conduct resilient operations in space (with or without a partner). The second was beliefs regarding the importance of different measures of resilience; for instance, whether the ability to avoid an attack is more important than the ability to recover quickly from an attack. The final survey instrument is documented in Appendix D.

We then applied the survey to a small sample of SMEs. The goal of the survey was to gather sufficient data to determine the independence of factors to be included in further quantitative analyses regarding the use of partnerships to build operational resilience for the MILSATCOM mission. An example of just such a quantitative analysis is described in Appendix B. The factors analysis for several of the more noteworthy paired logic models is discussed above to illustrate how we assessed independence and possible confounding effects.

In the next chapter, we document the overall lessons learned from applying the methodology.

Chapter 3. Application of the Methodology

In this chapter, we review the results from applying the methodology and describe what we learned from our analysis of the responses to the community attitudes survey.

What We Learned About Independent Factors

A summary of what we learned about the independence of our hypothesized dimensions of resilience is provided in Table 3.1. Overall, of the 18 hypothesized independent factors (nine pairs), our survey provided evidence of independence for only six factors (three pairs). Of those, in our subsequent analysis, we ultimately decided not to model the integration of augmenting versus core resources as independent because we strongly suspect that lack of trust in the United States' ability to supply *sufficient* resources might have confounded this portion of the survey. The two remaining pairs that are independent are associated with interoperable situational awareness and integrated planning and resource allocation.

Table 3.1. Factors Analysis Results

Dimension of Resilience	Paired Logic Models	What We Learned from Factors Analysis
Dependency	• Fighting in a coalition creates opportunities to confound adversary decisionmaking • Fighting in a coalition creates dependencies on partners that constrain national decisionmaking	**Independence cannot be assumed**. These are mirrored and perhaps correlated. We suspect that U.S. doctrine regarding the need to fight as a coalition is the actual independent variable, not the opportunities or constraints that arise from that decision.
Interoperable SA	• Shared situational awareness from multiple sources ensures that commanders have the information they need to make decisions • Local situational awareness that does not rely on external data ensures that commanders have reliable information, allowing rapid decisionmaking	**These might be independent factors**. Therefore, multiplicity of sources for SA and the reliability of those sources are modeled independently in later analyses.[a]
Planning and resource allocation	• Partnerships that have clear boundaries but separate planning can respond quickly, leading to more resilient operations • Partnerships that are fully meshed with integrated planning provide a clear understanding of total system capability, leading to more resilient operations	**These might be independent factors**. Therefore, in later modeling, the directionality of organizational ties is modeled separately from the strength of the tie (i.e., the degree of integration).

Dimension of Resilience	Paired Logic Models	What We Learned from Factors Analysis
Integration: augmentation or core capability	• To assure resilient space operations, the United States should supply all of the core capabilities needed for operations • To assure resilient space operations, the United States should augment core capabilities with those from partners	**These might be independent factors.** However, we strongly suspect that trust in U.S. ability to supply sufficient resources are confounding the results of the survey, and we did not include this dimension in our later analyses.
Integration: tailored or full-spectrum capability	• Operational resilience can best be achieved if partners bring fully capable systems that are interchangeable with U.S. systems • Operational resilience is best achieved if partners bring less capable but interoperable systems	**Independence cannot be assumed.** These might be correlated. The orthogonality we hypothesized (degraded trust of less capable systems) also does not appear to be a factor. Therefore, the capability of the systems is not modeled in our resilience analyses.
Integration: diversity or proliferation	• Minimizing the diversity of resources simplifies the coalition's ability to reallocate resources and achieve resilience in operationally relevant time frames • Adding greater diversity of resources complicates the adversary's decisionmaking, resulting in a more resilient architecture	**Independence cannot be assumed.** These are mirrored and perhaps correlated. We suspect that attitudes about the importance of robustness versus responsiveness to adversary attack is the actual independent variable.
Doctrine as an enabler	• Having similar doctrines leads to stronger partnerships and more resilient operations • Different but synergistic doctrines can be leveraged to provide greater freedom of action and more resilient operations	**Independence cannot be assumed.** These might be correlated. However, we did find evidence that commercial partners had a different distribution than military respondents. We hypothesize this might be because of differences in perception about the value of freedom of action.
Organization as an enabler	• A more loosely integrated command and control system is more operationally responsive • Fully integrated command and control systems allow planners to use all systems to best effect, resulting in a more robust architecture	**Independence cannot be assumed.** These are mirrored and perhaps correlated. We suspect that attitudes about the importance of robustness versus responsiveness to adversary attack is the actual independent variable.
Training as an enabler	• Repeated joint exercises with partners over many years are the best way to ensure effective partnerships in the field • Immediate pre- or post-engagement joint exercises with partners are the best way to ensure effective partnerships in the field	**Independence cannot be assumed.** These might be correlated. It is the joint exercise itself (existence of a tie), not its recency or repetition, that should be modeled.

[a] Note that today's consensus from the community surveyed is that multiplicity of sources for SA is viewed positively, while the reliability of local SA is judged more neutrally. We suspect that had we surveyed the fighter pilot community, we might have gotten a very different consensus on these statements. The reasons for this suspicion are two-fold: (1) Fighter pilots must make decisions about whether and how to engage an adversary on a much shorter timeline than spacecraft operators, and (2) all SA about space is in some sense *remote*, making the distinction between the reliability of local versus external sources of SA immaterial.

What We Learned About Trust

Recall that we asked about trust in two ways. The first used a set of statements to determine if U.S. military personnel trust the current system to produce the materiel, training, tools, and procedures needed for resilient MILSATCOM operations. The answer was a distinct "no." Response statistics are provided in Table 3.2. The statement that obtained the most agreement was that "the U.S. has the diversity of resources needed to achieve robust operations." Response to this question was distinctly bimodal: 47 percent of respondents disagreed, but 33 percent agreed. The other three statements each received only 13 percent agreement and no strong agreement.[22] The statement that respondents most strongly disagreed with was that "the U.S. has the tools and procedures needed to reallocate resources to achieve resilience in operationally relevant timeframes." Our respondents appear to be sending a very clear message: **Although the United States needs greater numbers (and perhaps diversity) of MILSATCOM resources, there is an immediate need to design, deploy, and train operations personnel in the use of tools and processes to rapidly reallocate the resources that the United States already has.**

Table 3.2. Responses to Statements Regarding Trust in the System

Trust in System Statement	Response
The United States has enough resources to achieve robust operations, independent of partners.	13% of respondents agreed; none strongly agreed
The United States has the diversity of resources needed to achieve robust operations, independent of partners.	Although most respondents disagreed, the overall distribution is bimodal: 47% disagreed and 33% agreed
The United States has the tools and procedures needed to reallocate resources to achieve resilience in operationally relevant time frames.	13% of respondents agreed; none strongly agreed
The United States has the tools and procedures needed to incorporate diverse resources into planning, thus enabling a more robust architecture.	13% of respondents agreed; none strongly agreed

As we noted in Chapter 2, if our respondents do not trust that they will have the tools and training to effectively plan and utilize capabilities, then integrating coalition and commercial partners into space operations is unlikely to be seen as a benefit independent of what the partnership brings to the warfight. This might explain the lack of correlation we found regarding our questions about diversity versus proliferation of materiel and the capability of materiel brought by partners.

Our second line in inquiry regarding trust was about the trust between the United States and potential partners. We found virtually no differentiation as to the type of partner or the direction

---

[22] Recall that we intentionally biased our statements to be positive and to nudge respondents toward a positive sentiment. Under these circumstances, a 13-percent agreement is a resounding rejection.

of trust. Partners are generally trusted, with the commercial partners being deemed perhaps slightly more trustworthy than military partners.

What We Learned About Sensitivity to Operational Vignettes

Overall, the operational vignettes appear to have significantly shifted the distribution of response in only three of 12 cases. Interestingly, in all three cases, the distribution shifted to bimodal; that is, where it had an effect, the more constrained scenario appears to change sentiments and, perhaps, challenge conventional assumptions. The three statements are

1. Operational resilience can best be achieved if partners bring fully capable systems that are interchangeable with U.S. systems.

   Some of the respondents who viewed the more constrained vignette disagreed with this statement; of the respondents who did not view the second vignette, none disagreed. It appears that respondents who viewed the more explicit vignette describing the hosted payload partnership were more willing to see the value of less capable systems.

2. To assure resilient space operations, the United States should supply all of the core capabilities needed for operations.

   Viewing the more constrained vignette appears to have increased the level of disagreement with this statement (i.e., more strong disagreement and less neutrality). Again, respondents who viewed the description of the hosted payload partnership were more willing to acknowledge that the U.S. military does not need to be fully self-sufficient with respect to core capabilities.

3. Partnerships that have clear boundaries but separate planning can respond quickly leading to more resilient operations.

   While respondents who did not view the constrained vignette were primarily neutral (45 percent) on this statement, respondents who viewed the vignette took sides: they either agreed (64 percent) or disagreed (27 percent), with few being neutral.[23]

These observations leave us with questions that we were unable to answer within the time and budget constraints of our study. These questions include

- What was polarizing about the vignette regarding the separate planning statement? Is it the specificity, the type of partnership we chose, or something else?
- Why did viewing the vignette only significantly impact response to one of the paired statements about a dimension of resilience while having little impact on the response to the paired statement?

---

[23] This could be interpreted as a fairly clear endorsement of the hosted payload partnership arrangement detailed in our vignette. This should not, however, be seen as an endorsement of *all* hosted payload partnerships.

What We Learned About Resilience

Recall from Chapter 2 that our survey included a section that asked respondents to rank order the measures of resilience in order of agreement. The results of that rank order are provided in Table 2.5. This inquiry was placed at the beginning of the survey, and our analysis (documented in Chapter 2) indicates that robustness (the ability to avoid or minimize service impacts of an attack) is valued more highly than responsiveness (the ability to recover quickly or to fine tune a response to an attack).

However, this was not our only inquiry about resilience. At the completion of the survey, we asked respondents to agree or disagree with statements regarding the importance of each measure of resilience. Most respondents did not differentiate but simply agreed that all were equally important. For those that did differentiate their degree of agreement, half were consistent with their original ranking and half were inconsistent. Whether a respondent was consistent or inconsistent with their original ranking does not correlate to whether they viewed the additional hosted payload vignette. Therefore, this second inquiry regarding preference for specific resilience measures yields no insight.

What We Learned About Partnerships

Other sets of ranked statements were designed to provide insight into how our respondents think about the value of joint training and the factors that lead to successful partnerships. The ranked order results indicate that our respondents value the joint understanding and informal bonds formed in joint training more highly than any improvements in expertise that might arise from the training. Transparent objectives and competence of a potential partner are more important than commitment to a partner's interests. While these results should inform future USSF exercises conducted jointly with partners, it is our observation that these lessons are already incorporated in planning for such events as the annual Schriever Wargame and Space Flag.[24]

Recommendations for Improving Measurements of Community Attitudes

During survey preparation, we had significant discussions as to whether the three-part nature of our logic model (input-output-outcome) was too complex to provide a good signal regarding community attitudes. We wanted to learn about outcomes (resilience measures), but we were concerned that by providing both an output and an outcome, we would be unable to distinguish whether our respondents agreed or disagreed with the logic that inputs lead to outputs versus that outputs lead to specific measures of resilience. Ultimately, we decided to state most outcomes simply as leading to "more resilient operations" or "robust operations" and to query separately

---

[24] This observation is based on RAND researcher participation in these events.

regarding attitudes about the measures of resilience (avoid, absorb, and recover from attacks). Given the lack of differentiation provided by our respondents to our queries about the importance of the different measures of resilience, we believe that including a generalized outcome statement was appropriate and would do so again.

In terms of whether to use a Likert scale or a ranked-order approach to make inquiries on a specific topic, our only caution is that each method has its place in research design.[25] A Likert scale does not force respondents to choose whether they agree more with *A* than *B*, but it does provide insight as to the relative strength of agreement. Meanwhile, a ranked ordering forces respondents to choose which response they agree with more, but it does not give insight as to whether *A* engenders a much stronger agreement than *B* or only slightly more agreement. Use of the Likert scale is essential to factor analysis, a technique that relies on the distribution of relative strength of agreement to determine the independence of factors. However, if our goal had been to elicit a consensus position regarding a specific topic of interest, we find the ranked order voting provides objective evidence despite the small number of respondents to our survey.[26]

We also have rereviewed our vignettes—especially the second vignette describing the hosted payload partnership specifics—for language that might be prejudicial. We did not find such language. Instead, we believe that it is the greater specificity of the vignette that causes our respondents to answer with a wider variety of sentiments. Where a lack of specifics in a vignette allows respondents to bring their own experience (and biases) to their answers, specifics might force them beyond their preconceptions and to confront conventional wisdom.

---

[25] For more information on these two types of survey designs, see Kathryn Phillips, F. Reed Johnson, and Tara Maddala, "Measuring What People Value: A Comparison of 'Attitude' and 'Preference' Surveys," *Health Services Research*, Vol. 37, No. 6, December 2002.

[26] Using a Likert scale to obtain statistically significant insight as to a consensus position would require a much larger sample size than we had for our survey.

Chapter 4. Summary and Conclusions

Throughout Chapter 3, and to a lesser extent Chapter 2, we expounded on the multiple lessons we learned regarding both the methodology itself and the application of the methodology to the topic of integrating coalition and commercial partners into the MILSATCOM mission. In this chapter, we summarize the most important of those lessons.

Reflections on the Methodology

Overall, our experience causes us to question the generalizability of quantitative research results based on SME consensus, especially for a topic as ambiguous as *resilience*. Such consensus appears to be deeply affected, not so much by time and operational context, as originally hypothesized, but—most importantly—by confounding factors.[27] The primary confounding factor we found in our exploration of community attitudes is that U.S. military personnel overwhelming believe that the United States lacks the tools, training, and procedures needed to rapidly reallocate MILSATCOM resources in response to an attack. Until this is remedied, more-detailed analyses regarding how best to integrate coalition and commercial partner resources may be biased by a lack of trust that those resources can and will be properly integrated. For this reason, our quantitative analysis, although documented in Appendix B, is not used to generate recommendations.

Our methodology does appear to properly identify independent factors to be used in resilience analyses. We believe our survey design is properly constructed for factors analysis and exploring the independence of factors more generally.[28] Much of the survey's power derives from grouping the paired statements to nudge respondents to view the statements as dependent. If respondents instead view the statements as independent, this provides stronger evidence of independence than if respondents had viewed the statements individually.

However, we were unable to demonstrate that we could measure factors that change with operational context: Operational context noticeably affected SME response in only three of 12

---

[27] A confounding factor is one that is not considered during hypothesis (cause and effect) formation but that fully impacts both cause and effect in a way that invalidates the hypothesis. A simple example of a confounding factor is as follows: Suppose you are exploring the relationship between shark attacks and ice cream sales and find that shark attacks do indeed precede an uptick in ice cream sales. The confounding variable here is temperature. There are more sharks in warm water and more ice cream sales in warm weather.

[28] We do have one caution regarding this use of the survey results. We had to continually remind ourselves that the purpose of using the Likert scales and paired logic model statements was *not* to establish a quantitative figure of merit but to evaluate the independence of factors. Yet, because we had hard numbers, the temptation to use them as a figure of merit was very strong, even though we knew our sample size was too small for the numbers to be significant in many cases.

cases. In terms of setting the operational context, we note that a lack of specifics in a vignette allows respondents to bring their own experience (and biases) to their answers. In contrast, a more specific vignette appears to force them beyond their preconceptions and to confront conventional wisdom.

Overarching Conclusion and Recommendations

The broad conclusion from our research is that the baseline resilience of USSF MILSATCOM is inadequate. This conclusion is based on our interviewees who told us the following:

- The USSF currently does not have sufficient staffing nor adequate tools to rapidly reallocate MILSATCOM resources among users, which severely limits the resilience of the overall MILSATCOM architecture. Although the staff can and will do some reallocation in a very limited crisis (e.g., loss of a single satellite because of on-orbit failure), USSF is currently staffed and trained to address the deliberate planning process.
- The tools used in planning and allocation are stovepiped among the various MILSATCOM constellations. This lack of integration severely hampers reallocation of users among the various constellations.
- Although integrating coalition and commercial partners offers the promise of increased overall architecture resilience, the specific nature of the formal agreements and contracts currently limits the fungibility of these assets among users. Furthermore, the disparate planning and allocation processes and tools would make any integration attempt with MILSATCOM assets a very ad hoc undertaking, especially given the current state of MILSATCOM planning and allocation.

This conclusion is bolstered by analysis of the community attitudes measurement survey responses:

- Respondents were nearly unified in their rejection of the statement that the USSF has sufficient MILSATCOM resources to be resilient in the face of the concerted adversary jamming attack vignette that we provided as operational context for the survey. However, there is less consensus as to whether the USSF has sufficient diversity of resources.
- Respondents overwhelmingly rejected the statement that the USSF has the tools and processes to integrate coalition and commercial partners into MILSATCOM operations.
- Respondents also overwhelmingly rejected the statement that the USSF has the tools and processes needed to reallocate resources (whether alone or with partners) in an operationally relevant time frame.

Although all the above observations must be addressed, we recommend that USSF first prioritize the development of tools and processes capable of reallocating MILSATCOM resources in an operationally relevant time frame and train operations personnel in their use. Without those tools and processes, the utility of adding resources, whether through partners or through USSF acquisitions, is limited.

Appendix A. Use Cases from Public-Private Partnerships

In this appendix, we review different public-private partnership models and the levels of control that they could offer to the USSF. We then document two such partnerships, with an eye toward formulating lessons learned.

Public-Private Partnership Models

There are many ways in which the U.S. government (USG) can form a partnership with private companies. The following have been used in the past for military SATCOM:

- **Satellite lease:** USG leases entire commercial communication satellites designed exclusively to meet their requirements. The private company finances the design, integration, and launch of the satellite and, sometimes, its operation. USG has exclusive access to the on-board communications bandwidth, coverage, and services of the satellite (or satellite constellation), and the commercial company has guaranteed revenues for the life of the satellite or constellation or for a fixed contractual period of time.
- **Anchor tenant:** USG signs on as *anchor tenant* to a commercial communications satellite designed for general commercial use. The private company finances design, integration, launch, and operations but has less business risk because USG guarantees a minimum service usage for a fixed period. Service above the specified minimum level of usage can be provided at (current) market price.
- **Transponder lease:** USG can also lease communication transponders on a commercial satellite at market price. Although USG is just one more customer of the commercial provider, USG does not have to compete with other customers for access to *their* transponders: The bandwidth is guaranteed to be available when it is needed. The downside of this approach is that costs are fixed whether or not the capacity is used.
- **Service level agreement (SLA):** USG can also acquire services on a commercial satellite system through an SLA. As with the transponder lease, USG is simply one more customer of the commercial provider. Unlike with the transponder lease, USG must compete with other customers for the use of the satellite's bandwidth, though some of that bandwidth might be guaranteed by the terms of the SLA.[29] In this case, USG is buying a service and only pays for bandwidth used.
- **Hosted payload:** USG asks a private partner to host a government-specified payload on a commercial satellite. Typically, USG pays up front for the communications payload and its integration into the commercial system. Integration costs might include provisions to amortize the payload use of satellite resources, such as electrical power, communication links, heat, or thermal cooling over the life of the satellite. The hosted payload approach allows USG to maintain secrecy regarding specific features, vulnerabilities, and

---

[29] An SLA might include any number of incentives or penalties to ensure the provider meets a specified quality of service. Depending on how the SLA is written, service under combat conditions might not be included.

operations of their communications payload while sharing the costs of satellite resources, launch, and basic housekeeping operations with commercial payloads.[30]

In the following section, we examine operational factors that might impact modeling of these different partnership constructs.

Level of USSF Operational Control

These public-private partnerships can use a variety of operational concepts and organizational roles to manage both the operational health and status of the satellites themselves and the operation of the satellite payloads. In all cases (except perhaps with the leased satellite), the accountability of managing the health and status of the satellite itself is typically given to the commercial partner. It is the operation of the payloads—those elements on board the satellite that provide the communications service between DoD and coalition warfighting systems—where we see the most variability. Typical constructs are described below:

- **Partner controlled:** In the case where an SLA is used, configuration and monitoring of the satellite payload is done entirely by the commercial partner. DoD's interface is through a gateway that interacts with the payload or the partner's network operations center (or both) to obtain communications bandwidth in specific geographic areas or between specific warfighting systems. Specific measures might need to be in place to ensure operational security.
- **USSF controlled:** USSF operators at either a separate or collocated network control center directly monitor and configure the communications payload, often using dedicated links that are not under the control of the commercial provider. It is common in these cases for USG to contract with the commercial provider for a backup communications link to the payload for contingency use.
- **Hybrid control:** In many cases, the responsibility to monitor the health and safety of the communications payload remains with the commercial provider, while configuration of the communications bandwidth (geographic region coverage and connectivity between warfighting systems) is performed by USSF operators.

Use Case: Hosted IRIS Payload on Intelsat-14

The Internet Router in Space (IRIS) payload hosted on the Intelsat-14 communication satellite is a unique public-private partnership where DoD paid a commercial provider (Intelsat) to not simply a host a government payload but to integrate that payload into the provider's C- and Ku-band communications uplink and downlink. Although DoD has for years routed internet protocol (IP) packets across satellite links and Cisco had conducted early demonstrations of IP

---

[30] One hosted payload being contemplated by USG is to ask one or more commercial SATCOM companies to host Link 16 encrypted data links to connect about 30,000 airborne and ground-based U.S. military and NATO systems. See Sandra Erwin, "Space Force Thinking About NASA-Style Partnerships with Private Companies," *SpaceNews*, June 4, 2020.

routers in space, the payload hosted on Intelsat-14 was Cisco's first IP router built with radiation-hardened electronics and integrated to provide a mesh network of the satellites' operations uplink and downlink channels as a joint capability technology demonstrator project.[31] After its 2009 launch, DoD was able to demonstrate the value of onboard routing to regional users over most of North America, South America, and the Caribbean, as well as Western Europe and Western Africa through the Defense Information Systems Agency's (DISA's) SATCOM II General Service Administration services lease arrangements. The regional users included DoD's U.S. Strategic Command and Southern Command, as well as the Royal Netherlands Navy, U.S. Coast Guard, Joint Interagency Task Force South, and NATO Consultation, Command and Control Agency personnel.[32] The operational assessment period was February 1, 2010, through May 24, 2010.

Operational Benefits

A 2011 Cisco white paper touts the operational benefits of a fully meshed on-board network as providing intelligence information-sharing among edge users in net-centric operations to promote greater cooperation among joint, interagency, intergovernmental, and multinational partners toward meeting their counterterrorism mission needs.[33]

According to the white paper, the technical advantages of the IP-based networking include:

- 50 percent faster transmission times enabled by a "single-hop" integrated mesh configuration
- Onboard regeneration for a power reduction of 3 dB or more, which results in reducing the size of ground antennas
- Flexible bandwidth on demand with higher throughput than traditional systems, designed to support larger transmissions such as real-time video from unmanned aerial vehicles or access for underserved users
- Ability to interface with legacy systems to support connectivity across joint groups or forces and use an encrypted IP connection to protect classified information[34]

---

[31] The radiation-hardened electronics were built by SEAKR Engineering, and ViaSat provided a software-defined radio to integrate the system into Intelsat's communication links. Traditionally, satellite uplinks and downlinks are treated as a hub-and-spoke network rather than a fully meshed network.

[32] Similar demonstrations of a capability to fully mesh onboard routing for the commercial sector have failed to produce a thriving market. One such example was Eutelsat's 2004 Skyplex payload (see European Space Agency, "SkyPlex: Flexible Digital Satellite Telecommunications," webpage, March 9, 2004). Another was HughesNet's direct-to-home internet service via satellite. HughesNet's first satellite launched in 2007, and, unlike Eutelsat's Skyplex, the HughesNet service continues as a viable business.

[33] Cisco, *Improving Communications Effectiveness and Reducing Costs with Internet Routing in Space: A DoD Joint Technology Capabilities Demonstration*, white paper, January 2011.

[34] Cisco, 2011, p. 1.

During IRIS operational testing, Cisco surveyed the participating services and reported that

- 100 percent of respondents cited improvements in specific quality of service metrics, including call clarity, success rate for data transmission, and persistence of signal while in motion.
- 75 percent of respondents noted improvements in point-to-point message transmission times and their ability to meet deadlines for performance and near-real-time mission completion, with high availability.
- 75 percent to 80 percent of respondents agreed that IRIS provided better reliability, capability, capacity, and robustness in overall communications service compared to traditional systems.

Operational Utility Assessment

A 2010 independent Army Space and Missile Defense Command Battle Lab–generated IRIS Operational Utility Assessment (OUA) briefing covers summary-level assessments of operational issues.[35] The assessment includes two specific recommendations for future coalition partnerships that are pertinent to our study:

- the need for more detailed agreements for developing jurisdictional protocols for operational support during situations that require more than one foreign coalition partner to respond
- the need to develop a coalition doctrine for continuing the level of interaction needed by operational forces and for periodic exercises to test a combined doctrine.

From the OUA brief, we have identified a set of lessons learned segregated by DOTmLPF element as described in Table A.1. These lessons learned might be helpful to the USSF when procuring services from commercial satellite communications providers or making decisions regarding hosted secondary payloads.

---

[35] Mike Florio, "Joint Capability Technology Demonstration (JCTD): Internet Routing in Space (IRIS), Operational Utility Assessment (OUA)," briefing slides, U.S. Army Space and Missile Defense Battle Lab, July 31, 2010.

Table A.1. DOTmLPF Summary Lessons Learned from the IRIS Joint Capability Technology Demonstration

DOTmLPF Element	Operational Use of Commercial Satellite Communications Services	Hosting of Secondary Payload
Doctrine	None	None
Operations	None	JCTD process model validated the operational utility of hosted payloads by rapidly defining requirements
Training	A USSF hybrid military occupation rating that combines network and satellite communications specialties might be required if the USSF is to fully embrace network convergence	USSF personnel need familiarity and training with new commercial capabilities to meaningfully interface with industry and ensure USSF equities are included in commercial companies' business planning[a]
Materiel	USSF use of International Traffic in Arms Regulations (ITAR)–compliant commercial off-the-shelf equipment is an efficient way to extend collaboration and situational awareness to coalition multinational partners on short notice[b]	JCTD process successfully leveraged relatively mature payload technology funded by commercial stakeholders to address a near-term capability gap
Leadership	USSF needs to continue to embrace commercial capabilities for creating win-win business models and solutions partnering with industry	USSF should foster partnership with commercial industry to close military communications gaps
Personnel	Benefit of reliance on commercial transport would shift the burden of network management services to commercial industry and requires fewer USSF personnel to maintain and monitor DoD portions of the network	None
Facilities	None	None

NOTE: JCTD = Joint Capability Technology Demonstration.
[a] This lesson learned is based on the observation that a viable and robust commercial industry has yet to emerge for space-based IP routers, despite the operational usefulness of such a system to the U.S. military. A hosted military payload is one way to close the gap left by the commercial industry.
[b] In our interviews with allied partners, more than one commented that use of a shared commercial service was one of the fastest ways to obtain interoperability in times of need.

Use Case: Commercially Hosted Infrared Payload

An Air Force Research Laboratory (AFRL)–designed and Science Applications International Corporation (SAIC)–built imager, Commercially Hosted Infrared Payload (CHIRP), was hosted on the SES-2 commercial communication satellite for a September 2011 demonstration. The objective was to explore the viability of a wide field of view (WFOV) overhead persistent infrared payload. The payload was designed to operate on-orbit for 12 months. Operations were conducted from three ground sites, with USSF AFRL end users working shoulder-to-shoulder with commercial operators at two of the three sites. The spacecraft operations center was managed by SES operators responsible for monitoring bus operations (telemetry, tracking, and command and health monitoring) of the communication satellites. These operators were directly responsible for

- forwarding CHIRP payload commands through a secure means to the SES-2 host satellite

- activating the CHIRP mission sensor by utilizing additional SES-2 commercial communications transponders (leased through DISA)
- managing the use of at least one or more of the SES-2 host satellite's 24 commercial C-band transponders to provide downlink of the infrared sensor's raw data.

The mission operations center was operated by Orbital Sciences personnel (Orbital Sciences was the manufacturer and payload integrator for the SES-2 satellite) on behalf of the USSF AFRL end users. These operators were responsible for

- formulating the commands sent to the CHIRP sensors via the spacecraft operations center
- receiving the raw sensor data and archiving it
- providing USSF AFRL imagery analysts with access to the stored raw data in near–real time.

The mission analysis center was located at an SAIC-managed site (SAIC developed the sensor). Operations personnel at the analysis center directly supported the USSF AFRL end users by

- retrieving CHIRP raw data stored in the mission operations center archive via a secure connection
- verifying the performance of the sensor
- providing a repository for analyzed CHIRP sensor data and reports.

Operational Benefits

Reports indicate that the CHIRP program accomplished all its mission objectives and had its initial demonstration period extended a total of three times.[36] According to the Air Force, there were more than 300 terabytes of WFOV sensor data collected by CHIRP, which enabled "analysis of more than 70 missile- and rocket-launch events and more than 150 other infrared events."[37] CHIRP was decommissioned in December 2013, following 27 successful months of demonstration.[38]

Lessons Learned Operational Assessment

We were unable to obtain an official operational assessment report for CHIRP but assembled a set of lessons organized by DOTmLPF-P from our research, presented in Table A.2.

---

[36] Government Satellite Report, "PODCAST: CHIRP Team Discusses Program and Benefits of Hosted Payloads," webpage, April 15, 2015.

[37] Government Satellite Report, 2015.

[38] "Air Force Discontinues CHIRP Mission, Cites Budget Constraint," Inside Defense, December 6, 2013.

Table A.2. DOTmLPF Summary of Lessons Learned from CHIRP

Elements	Military Operational Use of Commercial SATCOM	Hosting of Secondary Payload
Doctrine	DoD (USSF) must ensure IA *best practice* certification safeguards are in place for commercial or coalition nation host satellite ground processing operations.	DoD (USSF) must ensure IA certifications are in place for safeguarding the secure and protected transmission of classified military sensor secondary payload data from the host satellite to their ground operations centers.
Operations	In general, formally define roles, responsibilities, accountabilities, and authorities for each participating host satellite and USSF end-user organization. Assess the operational benefits (and costs) up front of dependence on contractors vs. USG or USSF operational roles, especially within secondary payload's DoD classified guidelines for transmitting, archiving, or processing secondary payload raw sensor data areas.	
Training	A USSF hybrid military specialty rating could improve USSF knowledge of commercial space operations tools and capabilities. This would be essential if USSF personnel were to need take over critical contractor roles.	
Materiel	Invest in ensuring or, if necessary, upgrading on-board NSA encryption or other approved space-based devices to protect classified sensor data.	Assess secondary payload design for functional dependence on host satellite (e.g., power, communications uplinks and downlinks). Ensure these dependencies are guaranteed by the host satellite partnership agreement.
Leadership	None	None
Personnel	None	None
Facilities	If USSF payload partnership is with coalition nation's host satellite, ensure that all operations sites on foreign soil comply with ITAR, COMSEC, and local foreign laws.	

NOTE: COMSEC = communications security; IA = information assurance; NSA = National Security Agency.

Appendix B. Network Analysis

We believe that until the United States remedies the lack of tools, training, and procedures needed to rapidly reallocate MILSATCOM missions across existing resources, more-detailed analyses regarding how best to integrate coalition and commercial partner resources will be obscured by a lack of trust that those resources can and will be properly integrated. However, given that caveat, our network graph analyses described in this appendix confirms the intuitions of our survey respondents, especially with respect to the value of joint training and exercises and the value of informal ties.

Understanding Partnership Impacts on Resilience

Modeling Partnerships with Network Analysis

Network analysis is one way to illustrate the connections between partners by which resources might be shared. Network analysis, however, is only one way to model resilience impacts of partnerships. Not all the independent factors we found can be illuminated by this type of analysis; for the factors that can be, this appendix describes our modeling and offers observations gleaned from the analysis.

Network analysis uses a graph of *nodes* and their connecting *ties* (i.e., the relationships between the nodes) to reason about the network's resilience or other attributes. In our case, the nodes of a network represent the DOTmLPF-P resources each party brings to the partnership. The size of each node and the direction and strength of the ties between nodes defines the network. To represent the strength of the relationship between two adjacent nodes in the network, we assign an edge strength between 0 and 1 for each such connection. To further represent the nature of the connection between nodes, we assign an edge sign (+1, -1). Networks with edges that can be assigned positive or negative values that represent the degree of connectivity between two nodes are called weighted signed networks.[39] The assignment of signs to these edges has been used frequently in the field of psychology to represent the relationships between people from different social groups. Here the edge sign is used to describe the nature of this relationship and a negative value represents distrust, dislike, or disagreement.[40] A negative weight in a weighted signed network can also indicate dissimilarity and, in general, can be used

---

[39] Srijan Kumar, Francesca Spezzano, V. S. Subrahmanian, and Christos Faloutsos, "Edge Weight Prediction in Weighted Signed Networks," *2016 IEEE 16th International Conference on Data Mining (ICDM)*, 2016.

[40] Nejat Arinik, Rosa Figueiredo, Vincent Labatut, "Signed Graph Analysis for the Interpretation of Voting Behavior," *International Conference on Knowledge Technologies and Data-driven Business (i-KNOW)*, October 2017.

to reflect whether the node connected via this edge is a contributing or detracting factor to the node it is connecting to.

By modeling a partnership relationship as a graph, we can leverage metrics from network analysis (such as centrality, graph density, shortest path) to examine how different relationships impact the overall resilience of the partnership.[41] Specifically, for this analysis, we built a representation of both the USSF and a proposed partner, modeling each party as being composed of nodes that represent each element of DOTmLPF-P. The ties between a party's own nodes represent their internal structure.[42] Ties that connect a node in one party's network to the other party's network represent how resources are shared and the bonds (formal or informal) that tie together these relationships. These ties might be heavily affected by trust. Using this basic graph structure, we can represent different types of partnerships and varying levels of trust between partners (and in one case that we will describe later, the level of trust in the USSF's ability to supply the tools and processes needed to effectively integrate partners into operations).[43]

Appendix A contains descriptions of various types of partnership arrangements the USSF could use in space operations. From these, we selected two partnership types and represented them as network graphs: (1) a minimally connected graph, shown in Figure B.1, representative of a hosted payload arrangement where the USSF operates the spacecraft payload and the commercial partner operates the spacecraft bus and (2) a more fully meshed graph, shown in Figure B.2, such as that which might someday exist between the USSF and a close ally.

---

[41] For a more in-depth introduction to network analysis, we recommend, for the clarity of writing and explanation of what can be a complex subject, Garry Robins, *Doing Social Network Research, Network-Based Research Design for Social Scientists*, SAGE Publications, 2015.

[42] We do not model the partners as having doctrine and policy nodes. This is not to imply that allied or commercial partners do not have doctrine or policy but only that the USSF has no influence over a third party's doctrine and policy.

[43] In the prior work that this analysis was inspired by (Dreyer et al., 2016), the authors developed a representation of the U.S. military space community as an arrangement of DOTmLPF-P nodes and used that network to reason about how U.S. investments in non–materiel means might impact resilience of U.S. space systems. At the time that work was performed (2015–2016), network analysis was quite immature. Analysis packages to compute what are now standard network statistics were in their infancy. Therefore, although we adopt the overall modeling approach from that work, we use standard network metrics to assess the resilience of the network. Furthermore, the goal of Dreyer et al. (2016) was to identify how portfolio investments impact resilience, but our goal is to generate insights into how the independent qualitative factors we found impact partnership dynamics and, thus, mission resilience.

Figure B.1. Hosted Payload Partnership

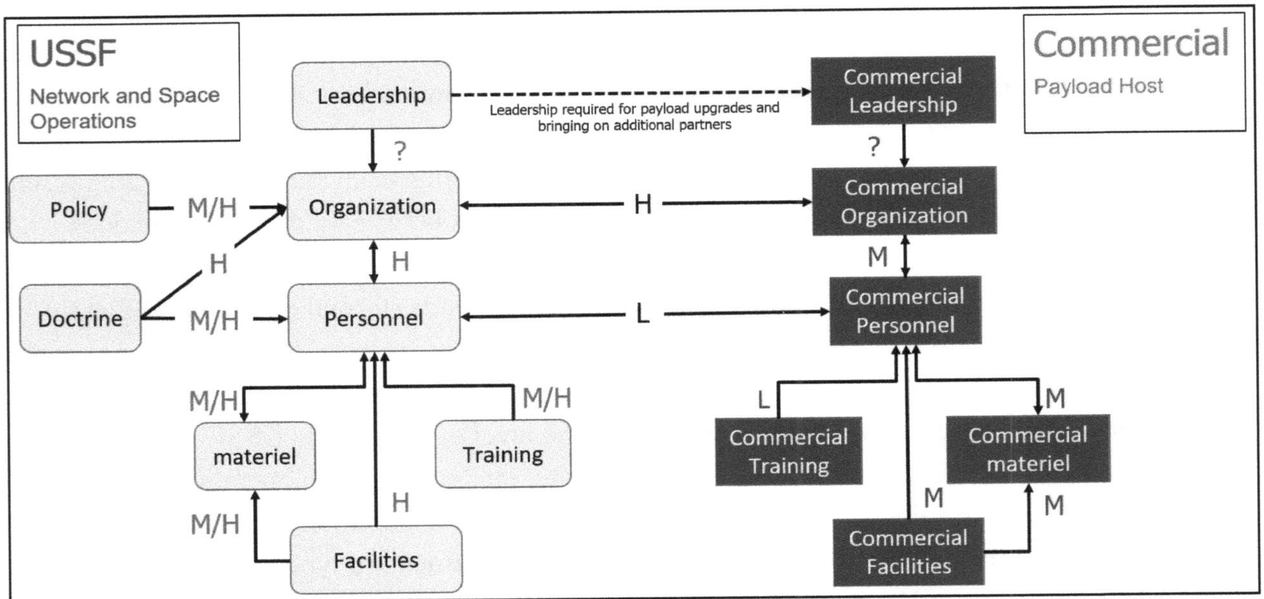

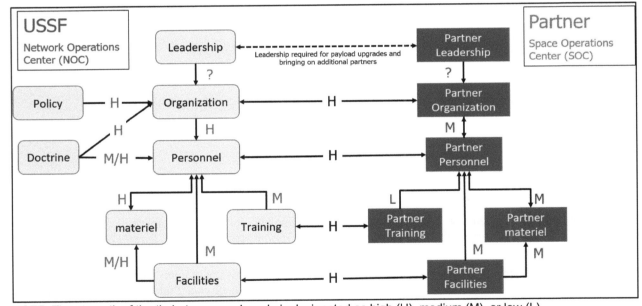

NOTE: The strength of the tie between each node is designated as high (H), medium (M), or low (L).

Figure B.2. Fully Meshed Partnership

NOTE: The strength of the tie between each node is designated as high (H), medium (M), or low (L).

Evaluating Resilience of the Mission and Partnership Dynamics

We leverage standard network analysis tools to measure the impact of varying the absence, direction, and strength of ties in the graphs shown in Figures B.1 and B.2. Note that the relationships between nodes are directional, as indicated by the arrows in the figures (i.e., in the terminology of network analysis, these are *directed graphs*).[44] Using a directional tie allows us to represent nonsymmetric relationships between nodes. The strength of the tie is designated as high (H), medium (M), or low (L) in the figures. Because these figures are only notional representations of the actual graphs used in our analysis, the relative size of each node has no meaning.[45] In the actual graphs, each tie (or *edge*) and each node are given a weight to describe the strength of the tie or the relative importance of the node, as shown in Table B.1.

Table B.1. Graph Weighting Parameters

Parameter	Definition	Measure
Edge Weight = w	Strength of influence between nodes	$w = (0, 1)$ 0 represents no influence between nodes 1 represents strongest influence between nodes
Node Weight = n	Relative importance of node to overall system resilience	$n = (0, 1)$ $\sum_{all\ n} n = 1$ 0 represents no impact on resilience 1 represents strongest impact on resilience

With the network graph defined, we can then begin to identify the impact of a partnership on the resilience of the network and, by extension, the mission. We begin by calculating the *strength of the shortest path* between each node. The model searches for the weighted shortest path between, for example, the USSF facility node and the partner's materiel node, and outputs each segment of the path and the number of segments.[46] The total strength of the influence is a product of edge weights for each segment.

$$j = influence\ of\ partnership\ on\ node = \prod_{i=path\ segment}^{all\ path\ segments} w_i$$

---

[44] A quick and reasonably complete tutorial of the difference between directed and undirected graphs can be found in Baeldung, "What Is the Difference Between a Directed and an Undirected Graph," webpage, November 24, 2022.

[45] In our later analysis of the impacts of the dimension of resilience, we do examine the sensitivity of the selected metrics of these graphs to the size of the material node.

[46] In Figure B.1, the USSF facility is only very indirectly connected to the hosted payload materiel (i.e., equipment). The shortest path is, in fact, quite long. For an event that occurs in the USSF facility to impact the partner's materiel, the event would have to first impact the USSF personnel either directly or through the USSF materiel. Then, the effect of the event on USSF personnel would have to impact the partner's personnel through the tie between personnel; we see that this is a low-strength tie for the hosted payload partnership. It might be that to impact the partner's personnel, the *weighted* shortest path is through the tie between the USSF and hosted payload organizations. Finally, the partner's personnel would need to take some action that impacts their materiel. Whereas, in Figure B.2, a more fully meshed partnership might have operators sharing a facility, forming a direct tie between the USSF and partner sections of the graph, and requiring only two steps for an event that occurred in the USSF portion of the facility to reach the partner's materiel.

The result is a measure of how the existence of the partnership impacts each DOTmLPF-P node.

Our next step measures the importance of each USSF node to the overall network resilience by summing, over USSF nodes, the node weight multiplied by the strength of the shortest path to reach that node from the analogous node in the partner network:

$$r = impact\ of\ partner\ network\ on\ USSF\ system\ resilience = \sum_{j}^{all\ nodes} j * n_j .$$

A shorter path should not be seen as either good or bad: More direct ties can improve mission resilience if they shorten a command path but can lower resilience if they also shorten the path a threat event would have to travel. To capture the increasing vulnerability to threat events (such as adversary attacks and system outages) that the partnership might bring, we identify whether there is a direct connection between analogous nodes in the graphs and assign a vulnerability score based on the strength of this direct connection, as follows:

$$v = total\ vulnerability\ introduced\ by\ partnership = \sum_{i}^{all\ node\ partners} c_i * w_i * n_i .$$

$$where\ c = 0\ if\ no\ direct\ connection,$$
$$= 1\ if\ direct\ connection$$

Another way to assess the vulnerabilities introduced to a mission through partnership is via the modularity metric. Modularity measures the strength of division of a network into groups (or *modules*): Here, our groups are defined as partner nodes and USSF nodes. Networks with high modularity have dense connections between the nodes within modules but sparse connections between nodes in different modules. A partnership graph that exhibits higher modularity is less interoperable, but the two parties can separate more easily to protect themselves if needed.

Additional metrics of network analysis are described in Table B.2 and will be used in the subsequent analysis of USSF partner networks.

Table B.2. Network Analysis Metrics

Metric	Definition
Betweenness centrality	Measures for each vertex the number of shortest paths that pass through this vertex
Graph density	Measure of connectedness of nodes in the graph
Modularity	Measures the strength of division of a network into groups
Resilience	Measure of support from the partner network via shortest path between partner and analogous USSF node, with the requirement that the path connects through the organization node for each network
Vulnerability	Measures the weighting of direct connections between partner and USG nodes
Resilience with informal network	Measure of support from the partner network via shortest path between partner and analogous USSF node

Partnership Modeling

For the next step in the methodology, we explore how the independent factors identified in the analysis above can be realized using partner network graphs. We then vary those factors and examine how they change the metrics of the graph. By determining the sensitivity to each factor, we build insights as to their relative impact on resilience outcomes.

What We Did: Representing Independent Factors in the Model

From our factors analysis of the community attitude survey results, we determined the independence of factors based on the distribution of responses. Table 3.1 summarizes that analysis. As noted in the last column of Table 3.1, these factors were found to impact only the DOTm aspects of the networks, and the mapping from dimensions of resilience to DOTm is not one-to-one. Therefore, Table B.3 maps from DOTm to dimension of resilience and then summarizes the graph variations used in our network analysis.

Table B.3. DOTmLPF-P Elements Varied to Examine Factors from Survey Responses

DOTmLPF-P element modeled	Dimension of Resilience	Graph Variations
Doctrine	Dependency, doctrine as enabler	• Shape of partnership network
Organization	Planning, organization as enabler	• Direction of the organizational tie between the USSF and partner networks • Strength of the organizational tie between the USSF and partner networks
Training	Training as enabler	• Existence of training tie between the USSF and partner networks[a]
Materiel	Interoperable C2/SA, Integration: augmentation vs. core	• Strength of the tie between USSF and partner materiel[b] • Size of partner materiel node • Size of USSF materiel node

[a] Only our fully meshed partnership graph has a direct tie between USSF and partner training. To isolate the effect of this tie, we analyzed the fully meshed network graph with the direct training tie present and then with it absent.
[b] Neither of our partnership graphs directly link USSF materiel to partner materiel. Therefore, to understand the impact of interoperable C2/SA, we varied the strength of the tie between the partner's personnel and their materiel.

We then conduct sensitivity studies to examine how the variations defined in Table B.3 impact the graph's metrics. For those factors impacting doctrine, we contrast metrics from the hosted payload partnership graph with those obtained from the fully meshed partnership graph, measuring the characteristics of the network structure itself. For those factors impacting organization, training, and materiel, we vary the existence, direction, or strength of the tie between the corresponding nodes of the USSF network and the partner network in one or both graphs. We score both the resilience and the vulnerability that the resulting partnerships introduce to the mission.

A limitation of our work is that we built our partnership graphs to reflect the trust and strength of paths for sharing between USSF and a potential partner given the form of the partnership. Our graphs do not explicitly model the speed of a response during a crisis; more nuanced factors, such as conducting exercises over time versus just in time; whether it is more optimal to have diverse resources internally versus more robust resources; or the differences between avoiding, absorbing, or quickly recovering from an attack. Although some of these could be studied using network graph techniques, it would require a different formulation of the graph. Although the goals of our research are met by providing an example of how to use the results of the community attitudes survey in a quantitative resilience analysis, an event tree analysis might be better suited to exploring some of these other factors.

DOTmLPF-P Elements in the Network Analysis Model

For the sensitivity studies described in this section, we primarily discuss impact using three metrics: vulnerability, resilience, and resilience with informal ties. We derived these above, and they are defined as:

- *vulnerability*: a measure that weights direct connections between partner and USSF nodes
- *resilience*: a measure of support from the partner network via shortest path between partner and analogous USSF node, with the requirement that the path connects through the organization node for each network
- *resilience with informal network*: a measure of support from the partner network via shortest path between partner and analogous USSF node; that is, support is not required to connect via the formal organization to organization tie

In Figures B.3 through B.7, all impacts are shown relative to a base condition, which is represented by the 0 location on the vertical axis. A 1 on this axis means either full vulnerability (the USSF is fully exposed to any degradation on the partner network) or full resilience (the USSF is fully insulated from all degradations of the partner network).

Doctrine

Doctrine is represented in the network structure as a separate node to allow us to vary the weight and significance of doctrine to USSF decisionmaking. However, how doctrine manifests in the partnership is reflected in the actual structure of each network graph. This is because doctrine frames how decisions are made and how resources are procured. The edge ties and weightings, along with the connection between nodes, reflect differences and similarities in doctrine between the USSF and the partner. Each of the subsequent discussions for various elements of DOTm will present results for both a hosted payload partnership structure and a fully meshed partner structure. These represent different doctrinal constructs regarding integration of partners and exhibit significantly different impacts on resilience.

The capture these characteristics, Table B.4 provides the graph density and modularity of each network graph. Density is a measure of the number of channels available to communicate and share resources among all nodes. As we will note in the upcoming discussion, the availability of these informal channels for decisionmaking and sharing is often quite impactful on resilience results. Modularity is a measure of the level of integration with the partner network. We see in our analysis that a higher level of integration corresponds to a higher availability of support from the partner network but also brings a higher vulnerability; incidents that occur in the partner network have more paths by which they might adversely impact USSF operations.

Table B.4. Summary of Network Structure Metrics

Metric	Hosted Payload	Fully Meshed Partner	Implications
Graph density	0.11	0.13	Higher graph density means a higher number of decisionmaking channels can be leveraged
Modularity	0.29	0.08	Lower modularity means more integration with partner but also more vulnerability because of connection with partner

Materiel

USSF and partner materiel in the model are configured to reflect both the sufficiency of resources and the availability of these resources. Recall that our survey results reveal a strong sentiment that the United States does not have sufficient resources to be resilient to adversary attacks. To reflect this in our model, the sufficiency measure of USSF materiel is decreased from 100 percent (fully sufficient) to 50 percent and then 10 percent. The impact of this degradation is shown in Figure B.3.

For this analysis, the base condition is that the USSF has 100 percent of the resources needed. We then plot the three metrics (vulnerability, resilience, and resilience with informal networks) to show the impact of integrating with partners as a function of the degraded sufficiency of USSF materiel. The metrics computed from the hosted payload graph are shown on the left, and the metrics of the fully meshed partner graph are shown on the right. As can be seen in the plots, either partnership structure adds resilience. The additional resilience is most significant with a fully meshed partner, but both types of partnerships bring the level of resilience above the initial USSF baseline.

Note that the impact of informal ties is limited for the hosted payload partnership—this is unsurprising given the structure of that partnership. However, for the more fully meshed partnership, the informal ties offer a significant improvement in relative resilience. This will be true for all the factors of resilience that we explore. Across all our sensitivity analyses, when comparing the two partnership models, we note that the fully meshed partnership offers only slightly better resilience with increased vulnerability. Unless the fully meshed partnership can leverage informal networks, the additional connectivity it brings might not be worth the risk.

Figure B.3. Impact of Insufficient USSF Materiel on System Resilience

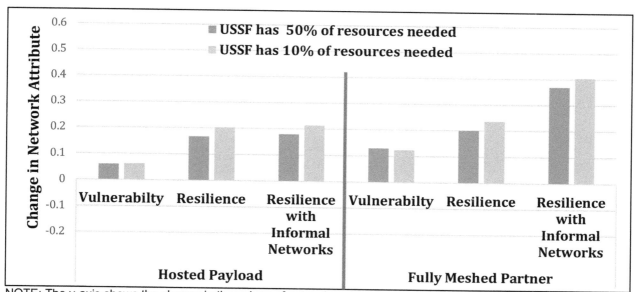

NOTE: The y-axis shows the change in the values of network metrics for vulnerability, resilience, and resilience with informal ties. Network attribute values range from 0 to 1, with 0 representing USSF with full materiel resources, and 1 representing either full vulnerability (the USSF is fully exposed to any degradation on the partner network) or full resilience (the USSF is fully insulated from all degradations of the partner network).

Factors analysis indicates that multiplicity of sources for SA of adversary actions and the reliability of these sources should be modeled independently. Although multiplicity of sources is expected to have similar impact as the USSF sufficiency of resources modeled above, to model the reliability of sources, we vary the strength of the edge weighting from the partner materiel node. An increase represents an increase in the reliability of SA from the partner (or materiel more generally) and a decrease represents lower confidence in the reliability of information about the adversary's impact on a partner's resources.

The results from varying the strength of the partner's materiel tie are shown in Figure B.4. Here, the baseline case (0 on the Y axis) represents a USSF with insufficient materiel and no partners. The metrics show that even with degraded SA of partner materiel, partnership still offers an increase in resilience over that baseline (USSF system on its own). Perhaps more importantly, the hosted payload partnership is largely insensitive to variations in the degree of reliability of the partner's SA.

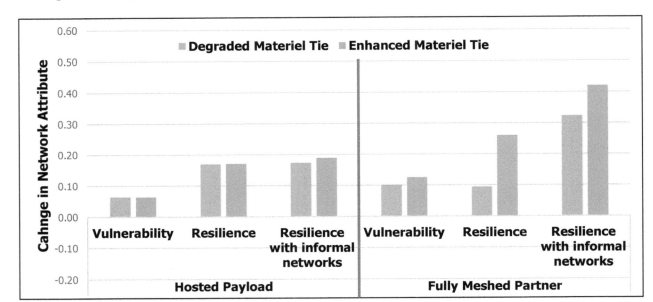

Figure B.4. Impact of Partner Situational Awareness Reliability on USSF System Resilience

NOTE: The y-axis shows the change in the values of network metrics for vulnerability, resilience, and resilience with informal ties. Network attribute values range from 0 to 1, with 0 representing USSF with no partners and insufficient materiel resources, and 1 representing either full vulnerability (the USSF is fully exposed to any degradation on the partner network) or full resilience (the USSF is fully insulated from all degradations of the partner network).

Our model is limited, however, to reflecting only the structural impacts of unreliable SA and materiel more generally. In the real world, unreliable SA or other partner materiel could have far-reaching consequences. Although the actual vulnerability score shown here is unchanged by the reliability of SA, if the adversary can exploit those vulnerabilities, the consequences could be substantial. This observation makes us hesitant to recommend the fully meshed partnership because of its relatively large vulnerability score.

Organization

From the factors analysis, we determined that planning and resource allocation might be independent factors. The ability to respond quickly should be considered a separate factor from the ability to present a clear understanding of total system capability. Each factor supports resilience in a different way. Although survey results indicate that partnerships with separate planning would enable higher resilience by facilitating quicker response times, our graphs do not include a time element, so we are unable to assess the impact of separate planning on responsiveness.

We can, however, assess the impact of integrated planning. To model this factor in our network graphs, we change the sign of the tie between the USSF organization node and the partner organization node. A negative sign for this tie indicates completely separate or perhaps even disjointed planning; a positive sign indicates fully meshed or integrated planning. In Figure B.5, we demonstrate how this change in the sign of the tie between the USSF and partner organization nodes impacts our two network graphs. In all cases, disjointed planning reduces the resilience of the USSF below that which it could achieve on its own, except for the case when informal networks are active in the fully meshed partnership. The fact that disjointed planning can reduce effectiveness should not be surprising. What is perhaps more surprising is the power of informal networks. Even in the hosted payload partnership, a negative tie between *formal* organizations can be greatly mitigated through *informal* networks. Integrated planning, on the other hand, improves USSF resilience in all cases, but it also introduces vulnerability, and that vulnerability is greater in the fully meshed partnership. Taken together, this observation might suggest that a hosted payload partnership with informal networks and some level of integrated planning is a reasonable way to balance resilience against vulnerability for the USSF MILSATCOM mission.

Figure B.5. Impact of Disjointed vs. Fully Integrated Planning on USSF System Resilience

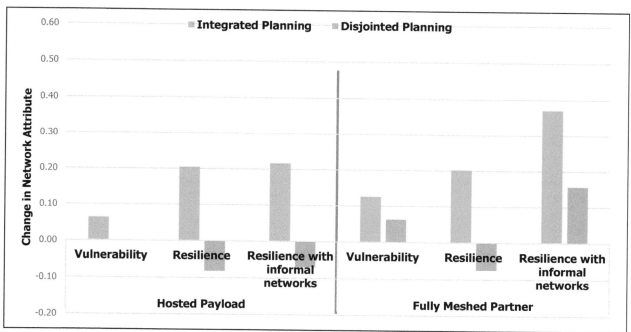

NOTE: The y-axis shows the change in the values of network metrics for vulnerability, resilience, and resilience with informal ties. Network attribute values range from 0 to 1, with 0 representing USSF with no partners and insufficient materiel resources, and 1 representing either full vulnerability (the USSF is fully exposed to any degradation on the partner network) or full resilience (the USSF is fully insulated from all degradations of the partner network).

To explore this hypothesis further, we vary the weight of a positive organizational tie between USSF and partner systems to reflect the degree of integrated planning (full, medium, and low) between them. The results of this analysis are shown in Figure B.6. This plot illustrates that even a weak organizational tie, representing less C2 integration, improves resilience while minimizing vulnerability. The greatest resilience—but also the greatest vulnerability—is achieved by creating a fully meshed partnership.

These results also demonstrate the nonlinearity in resilience if integration is confined to the formal organization-to-organization relationship. A hosted payload partnership with weak

51

planning but informal networks exhibits equivalent performance to a partnership with high integration but no informal networks. In the case of the fully meshed partnership, the lowest level of integrated planning with informal networks outperforms fully integrated planning if the partnership is constrained to operate only through formal organizational ties.[47]

Figure B.6. Impact of Organization Tie Strength on USSF System Resilience

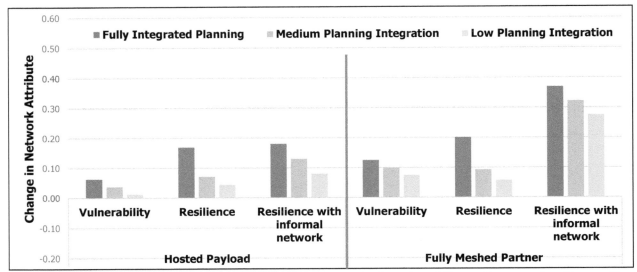

NOTE: The y-axis shows the change in the values of network metrics for vulnerability, resilience, and resilience with informal ties. Network attribute values range from 0 to 1, with 0 representing USSF with no partners and insufficient materiel resources, and 1 representing either full vulnerability (the USSF is fully exposed to any degradation on the partner network) or full resilience (the USSF is fully insulated from all degradations of the partner network).

Training

Our survey respondents indicated that building informal ties is one of the most important outcomes of integrated training. Furthermore, our factors analysis indicates that the *existence* of training, whether it be exercises over a long period of time or more timely (recent) exercises, is the most important factor to model. The role of training is significantly different for each partnership type we examine here. The hosted payload partnership graph does not have direct training ties between the USSF and their partner, and we assume that training is conducted separately within each organization. For the fully meshed partnership, however, there is a direct tie between the training nodes to indicate integrated training.

To understand the sensitivity of our resilience analysis to the existence of integrated training, we compare the network attributes of vulnerability, resilience, and resilience with informal networks with and without this direct training to training tie. Results, shown in Figure B.7, illustrate that integrated training provides a marked increase in resilience if informal channels

---

[47] We are not the first researchers to note the outsized impact of weak ties within a network. The seminal paper on this topic is Mark S. Granovetter, "The Strength of Weak Ties," *American Journal of Sociology*, Vol. 78, No. 6, May 1973.

can be leveraged for decisionmaking but provides a minimal difference if they cannot. Furthermore, introducing integrated training could concurrently introduce vulnerabilities. If informal networks cannot be leveraged for decisionmaking, the additional risk of conducting integrated training might not be warranted. In practice, this means exercises with allies and commercial partners that do not involve the actual operations personnel who will need to collaborate in combat—and that, therefore, do not build the informal networks needed—should be avoided.

Figure B.7. Impact of Integrated Exercises in a Fully Meshed Partnership

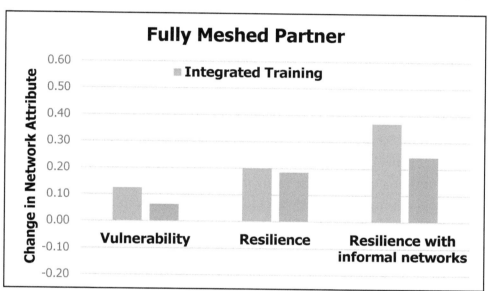

NOTE: The y-axis shows the change in the values of network metrics for vulnerability, resilience, and resilience with informal ties. Network attribute values range from 0 to 1, with 0 representing USSF with no partners and insufficient materiel resources, and 1 representing either full vulnerability (the USSF is fully exposed to any degradation on the partner network) or full resilience (the USSF is fully insulated from all degradations of the partner network).

What We Learned

Partnerships of both structures discussed here provide a level of resilience that surpasses the resilience USSF can achieve on its own, even if the partner network itself has unreliable resources. This might support a recommendation that the USSF address current resource insufficiencies by integrating coalition and commercial partners into operations rather than by procuring its own resources. However, introducing a partner network connection introduces vulnerability, and the operational environment and tolerance for risk must be considered.

Moreover, the absence of integrated planning between the organizations can lead to disjointed operations that eliminate the gains in resilience noted above. Even a low level of integration in planning is highly leveraged, especially if the partnership is not limited to the formal organizational communications channels.

When considering the type of partnership to pursue, our analysis shows that a hosted payload partnership introduces less vulnerability than a highly integrated partnership but also provides a smaller increase in USSF resilience. The latter is especially true if informal decisionmaking channels in a more highly integrated partnership can be leveraged. This is because informal channels are more influential than formal C2 integration or integrated training. Without these informal channels, a fully meshed partnership offers a minimal increase in resilience that might not outweigh the increase in vulnerability versus the hosted payload partnership.

Overall, the sensitivity analyses documented here lead us to conclude that if the USSF is to successfully capture resilience benefits offered by a partnership, one of two things must be true: (1) planning must be integrated to some level, or (2) informal decisionmaking channels must be enabled. The best case is when both are true.

Recommendations for Improving Partnership Modeling

Our formulation of the network graphs cannot tell us everything about partnerships and resilience. For example, we observed from survey responses that separate planning is expected to increase responsiveness, but the network graphs demonstrate that if separate planning results in disjointed planning, resilience decreases. In this, our network model properly captures impacts of planning on the *robustness* of a response to adversary attack but not the *speed* of the response or recovery from attack. A separate analysis should be designed to explore the tradeoff between robustness and recovery as a function of the degree of planning integration.

The adversary has a say in how USSF partnerships impact resilience. A skilled adversary can exploit the increased connectivity that comes with a highly integrated partnership. Although the USSF may boost resilience using partnership, it might not be worth the risk. Our scores of vulnerability and resilience are not on the same consequence scale, and their magnitudes cannot be compared to provide insight on this issue. The analysis documented here can be used to understand the relative vulnerabilities or relative resilience of different partnerships, but it cannot compare vulnerability relative to resilience; a separate analysis should be designed to explore the tradeoff between vulnerability and resilience.[48]

---

[48] Although a value of 1 on each scale represents full vulnerability or full resilience, the consequence of full vulnerability might be orders of magnitude higher than the benefits of full resilience.

Appendix C. Data Collection—Interview Protocol

The following interview protocol was used to elicit logic models regarding resilience. Text in italics is provisional instruction for the interviewer to adjust the questioning in cases where interviewees appeared to need additional context in formulating their discussion.

Interview Protocol

Establishing Context:

1. In the U.S. military, we often differentiate between strategic, operational and tactical levels of decision making. *[Provide definition if needed.]* Where in that hierarchy has your experience generally influenced your thinking on the topic of integrating allies into U.S. military operations? How about commercial partners? *[This will let us know where we should expect to spend more time in the interview. It also may provide a heads up as to how concrete the prompts may need to be.]*
2. How many years of experience do you have in space systems? In resiliency planning? In allied or commercial partnering with the U.S. military?

Part 1—Strategic

1. The National Defense Space Strategy released in June lays out the strategic need to integrate allies and commercial partners into plans, operations, exercises, engagements.

   a. When discussing integrating allies and comm partners, what does that entail? How do you think about this? *[Least concrete, most open]*
   b. Do you think it is strategically beneficial overall to do this (integrating allies and partners)? Why or why not? *[Ask if a more concrete prompt is needed]*
   c. What is the right mix of allied and commercial partners for military satellite communications and why? How do you weigh these advantages and disadvantages? *[Most concrete, bounds the discussion to MILSATCOM]*
   d. What are the disadvantages of integrating allies and commercial partners and how do you / could we mitigate them? *[Ask if they've only discussed advantages]*
   e. Are the strategic issues regarding integrating allies different than those for integrating commercial partners? *[Ask if they haven't differentiated]*

Part 2—Operational

2. A dimension in providing space domain mission assurance is resilience. Resilience has a number of sub-ordinate elements one of which is diversification.

   a. Can you describe diversification in your own words? How do you think about diversification from a mission assurance perspective? *[Least concrete, most open]*

b. What aspects of a space system should we diversify? Where does diversity provide the largest advantage? The least advantage? *[Ask if a more concrete prompt is needed]*

c. For military satellite communications, there are many things we could diversity. What types of diversity offer the largest advantages? Different orbits? Satellites? Designs? Ownership? The least advantage? *[Most concrete, bounds the discussion to MILSATCOM]*

d. What types of diversity are the hardest to achieve? *[Ask if they only talk about the good diversity brings]*

e. Do allies bring special advantages/disadvantages in diversity that commercial partners can't, or vice versa? *[Ask if they haven't differentiated]*

3. Integrating allies and commercial partners can create dependencies.

a. Can you talk about the dependencies that are created from allies and/or commercial partners from a mission assurance perspective? *[Least concrete, most open]*

b. What aspects of dependency on allies and/or commercial partners matter most to a decisionmaker at your level? *[More concrete]*

c. For the MILSATCOM mission, how might a dependency on an ally or commercial partner create opportunity? In what parts of the world? How about risk? *[Most concrete, bounded]*

d. What types of dependencies are the most advantageous to the U.S.? *[Ask if they only seem to talk about the disadvantages]*

e. Do dependencies on commercial partners create different opportunities and/or risks than dependencies on allies? *[Ask if they haven't differentiated]*

***Part 3—Tactical** [If we've had to bound the context to MILSATCOM in the above discussions, bound it for these questions also.]*

4. When thinking about diversity

a. What factors enable diversification?
b. What are barriers to achieving diversity?
c. How do you mitigate the barriers? How to prevent barriers?
d. What are the consequences if we can't mitigate *[barrier they mentioned]*? Does it change the probability of achieving mission success?

5. When thinking about dependencies upon allies and partners

a. What factors enable successful relationships (dependencies)?
b. What are barriers to achieving successful dependencies?
c. How do you mitigate the barriers?
d. What are the consequences if we can't mitigate *[barrier they mentioned]*? Does it change the probability of achieving mission success?

Appendix D. Data Collection—Community Attitude Survey

The following survey design was used to measure community attitudes. Text in italics provides instruction to the programmer who created the online tool. All surveys were administered electronically by RAND's Survey Research Group. The introductory material was provided to all respondents.

Introduction

This survey instrument is designed to measure community attitudes regarding whether and how integrating allies, coalition partners and commercial partners into the military satellite communications (MILSATCOM) mission enhances the resilience of military operations. In the first part of the survey, you will be asked to rank order statements related to this topic. In the second, you will be asked to rate on a scale from strongly disagree to strongly agree statements that presuppose how and why factors about the partnership build resilience.

Definitions

Some words used in this survey have multiple meanings. With the exception of the terms defined below, we ask that you use the definition commonly used in your area of expertise.

- **Doctrine**—a stated principle of government, military or corporate policy.
- **Partners**—when the word partner is used without a modifier it means any government, military or commercial entity that is engaged with the U.S. Department of Defense in providing SATCOM to warfighters.
- **Military**—when the word military is used without a modifier it means all DoD, allied and coalition partner organizations. It does not include the commercial entities that support the military.
- **Operations**—all activities involved in providing SATCOM to warfighters using currently deployed systems. As such, it includes developing doctrine, strategy, planning, and execution of tactics, techniques, and procedures for the SATCOM mission. It does not include the development of systems (i.e. satellite or terminal procurement).

Context

Please answer within the context of the following vignette:

U.S. service members and partners are collocated in theater, fighting in a coalition command structure. While different partners have brought their own SATCOM terminals and gateways, any nation's communication systems (including communication services procured from commercial providers) can be used to support the coalition within pre-negotiated prioritized limits. The coalition partners have trained together to respond to contingencies and work together in a combined SATCOM planning center to redistribute communications between their various systems in response to adversarial actions or rapid changes in the operational environment.

Currently the adversary is jamming several of the coalition's pre-planned communication channels. Multiple systems are available to fill the need for communications services, but replanning is necessary to establish new communication links that would circumvent the jamming attack.

Rank Order Questions

For each group of statements below, rank order them to indicate how strongly you agree. Use a 1 to indicate the one you most agree with, a 2 to indicate the statement you agree with next, etc.

1. Resilience for the SATCOM mission is best achieved through the ability to:

   a. _____ recover quickly after an adversarial action impacts service
   b. _____ avoid any service impact from adversarial action
   c. _____ minimize the service impact of adversarial action
   d. _____ finely tune the system in response to adversarial action

2. The most important outcome of joint exercises with partners is:

   a. _____ the informal bonds formed by training together.
   b. _____ a joint understanding of each party's tactics, techniques, and procedures.
   c. _____ the expertise gained by training together.

3. The success of a partnership is most dependent on both parties:

   a. _____ being transparent about their goals, objectives, capabilities, and constraints
   b. _____ believing the other has their interests at heart.
   c. _____ having the capability to effectively act in the joint interest.

Agreement Statements

Rate each statement below on a scale from strongly disagree to strongly agree.

This first set of questions focuses on the enablers of joint operations, including trust, doctrine, training, shared situational awareness, and concepts for command and control.

Note: only to be asked of U.S. DoD personnel:

1. The U.S. has enough resources to achieve robust operations, independent of partners.
2. The U.S. has the diversity of resources needed to achieve robust operations, independent of partners.

3. The U.S. has the tools and procedures needed to reallocate resources to achieve resilience in operationally relevant timeframes.
4. The U.S. has the tools and procedures needed to incorporate diverse resources into planning, thus enabling a more robust architecture.

Note: only to be asked of U.S. DoD and allied military personnel:

1. Coalition partners can rely on the U.S. for rapid response to overcome adversary threats.
2. The U.S. can rely on coalition partners for rapid response to overcome adversary threats.

Note: Ask of all survey takers (note to programmer—please keep the grouped statements on the same page/screen):

1. The military can trust commercial partners to strive to continue to provide service while overcoming adversary threats.
2. The military can trust commercial partners to provide robust services that can withstand attack.
3. Commercial partners can trust the military to defend them when they come under attack.

4. Fighting in a coalition creates opportunities to confound adversary decision making.
5. Fighting in a coalition creates dependencies on partners that constrain national decision making.

6. Having similar doctrines leads to stronger partnerships and more resilient operations.
7. Different but synergistic doctrines can be leveraged to provide greater freedom of action and more resilient operations.
8. Diverse thought on how to conduct operations is critical to achieving resilience.

9. Repeated joint exercises with partners over many years are the best way to ensure effective partnerships in the field.
10. Immediate pre- or post-engagement joint exercises with partners are the best way to ensure effective partnerships in the field.
11. A fully trained and capable partner is more critical than joint training.

12. Shared situational awareness from multiple sources ensures that commanders have the information they need to make decisions.
13. Local situational awareness that does not rely on external data ensures that commanders have reliable information, allowing rapid decision making.
14. A more loosely integrated command and control system is more operationally responsive.
15. Fully integrated command and control systems allow planners to use all systems to best effect, resulting in a more robust architecture.

16. Minimizing the diversity of resources simplifies the coalition's ability to reallocate resources and achieve resilience in operationally relevant timeframes.
17. Adding greater diversity of resources complicates the adversary's decision making resulting in a more resilient architecture.

NOTE: 50% of participants will be asked to answer the remaining questions given a revised vignette. The other 50% will continue with the original vignette.

Please rate the following statements assuming this additional information:

A military payload has been hosted on a commercial satellite constellation to provide robust coverage at minimal cost. The commercial provider manages the satellite constellation and operates several (nonmilitary) payloads. The USSF operates the military payload using an in-band communications link. If the in-band communication link to the military payload is disrupted or degraded, the commercial operator can utilize alternate communication links—on request and on a non-interference basis—to assist the military operations center with debug. For security reasons, connectivity between the military and commercial operations centers is limited to email, phone and secure file transfer.

This second set of questions focuses on the types of partnerships and the resources a partner brings to the warfight.

1. Partnerships that have clear boundaries, but separate planning can respond quickly leading to more resilient operations.
2. Partnerships that are fully meshed with integrated planning provide a clear understanding of total system capability leading to more resilient operations.
3. Shared strategic interests and priorities are necessary for productive partnerships.

4. To assure resilient space operations, the U.S. should supply all of the core capabilities needed for operations.
5. To assure resilient space operations, the U.S. should augment core capabilities with those from partners.

6. To assure resilient space operations, the U.S. should supply all elements needed for its PACE (Primary, Alternate, Contingency, Emergency) plans.
7. Resiliency of space operations can be improved by incorporating partner capabilities into PACE plans.
8. Including partner capabilities in space operations creates dependencies that confound adversary decision making.

9. Operational resilience can best be achieved if partners bring fully capable systems that are interchangeable with U.S. systems.
10. Operational resilience is best achieved if partners bring less capable, but interoperable systems.
11. If partner systems are less capable, it degrades the trust necessary for resilient operations.

12. Being able to avoid an attack is important to resilience.
13. Being able to absorb an attack is important to resilience.
14. Being able to recover quickly from an attack is important to resilience.

Abbreviations

AFRL	Air Force Research Laboratory
C2	command and control
CHIRP	Commercially Hosted Infrared Payload
DAF	Department of the Air Force
DISA	Defense Information Systems Agency
DoD	U.S. Department of Defense
DOTmLPF-P	doctrine, organization, training, materiel, leadership and education, personnel, facilities, and policy
IDCSP	Initial Defense Communications Satellite Program
Intelsat	International Telecommunications Satellite Consortium
IP	internet protocol
IRIS	Internet Router in Space
ITAR	International Traffic in Arms Regulations
MILSATCOM	military satellite communications
NATO	North Atlantic Treaty Organization
PAF	Project AIR FORCE
SA	situational awareness
SAIC	Science Applications International Corporation
SATCOM	satellite communications
SLA	service level agreement
SME	subject-matter expert
TTP	tactics, techniques, and procedures
USG	United States government
USSF	United States Space Force
WFOV	wide field of view

References

"Air Force Discontinues CHIRP Mission, Cites Budget Constraints," Inside Defense, December 6, 2013.

Arinik, Nejat, Rosa Figueiredo, and Vincent Labatut, "Signed Graph Analysis for the Interpretation of Voting Behavior," *International Conference on Knowledge Technologies and Data-Driven Business (i-KNOW)*, October 2017.

Baeldung, "What Is the Difference Between a Directed and an Undirected Graph," webpage, November 24, 2022. As of April 18, 2023: https://www.baeldung.com/cs/graphs-directed-vs-undirected-graph

Bobbitt, Zach, "What is a Confounding Variable? (Definition and Example)," webpage, Statology, February 19, 2021. As of August 2021: https://www.statology.org/confounding-variable

Brown, Timothy A., *Confirmatory Factor Analysis for Applied Research*, 2nd ed., Guilford Press, 2015.

Burch, Ron, *Resilient Space Systems Design: An Introduction*, CRC Press, 2019.

Center for Community Health and Development, *Community Tool Box, Chapter 2, Section 1: Developing a Logic Model or Theory of Change, Main Section*, University of Kansas, undated. As of August 2021: https://ctb.ku.edu/en/table-of-contents/overview/models-for-community-health-and-development/logic-model-development/main

Cisco, *Improving Communications Effectiveness and Reducing Costs with Internet Routing in Space: A DoD Joint Technology Capabilities Demonstration*, white paper, January 2011.

Dreyer, Paul, Krista Langeland, David Manheim, Gary McLeod, and George Nacouzi, *RAPAPORT (Resilience Assessment Process and Portfolio Option Reporting Tool): Background and Method*, RAND Corporation, RR-1169-AF, 2016. As of April 13, 2023: https://www.rand.org/pubs/research_reports/RR1169.html

Erwin, Sandra, "Space Force Thinking About NASA-Style Partnerships with Private Companies," *SpaceNews*, June 4, 2020.

European Space Agency, "SkyPlex: Flexible Digital Satellite Telecommunications," webpage, March 9, 2004. As of June 2021: http://www.esa.int/Applications/Telecommunications_Integrated_Applications/SkyPlex_Flexible_digital_satellite_telecommunications

Florio, Mike, "Joint Capability Technology Demonstration (JCTD): Internet Routing in Space (IRIS), Operational Utility Assessment (OUA)," briefing slides, U.S. Army Space and Missile Defense Battle Lab, July 31, 2010.

Glen, Stephanie, "Factor Analysis: Easy Definition," webpage, Statistics How To, undated. As of August 2021:
https://www.statisticshowto.com/factor-analysis

Government Satellite Report, "PODCAST: CHIRP Team Discusses Program and Benefits of Hosted Payloads," webpage, April 15, 2015. As of April 18, 2023:
https://ses-gs.com/govsat/defense-intelligence/podcast-chirp-team-discusses-program-and-benefits-of-hosted-payloads

Granovetter, Mark S., "The Strength of Weak Ties," *American Journal of Sociology,* Vol. 78, No. 6, May 1973.

Kennedy, John F., "Remarks on Signing Communications Satellite Act, 31 August 1962," Papers of John F. Kennedy, President's Office Files, August 31, 1962. As of July 2021:
https://www.jfklibrary.org/asset-viewer/archives/JFKPOF/039/JFKPOF-039-051

Krebs, Gunter D., "IDCSP—DSCS-1 (NATO 1)," webpage, Gunter's Space Page, undated. As of July 2021:
https://space.skyrocket.de/doc_sdat/idcsp.htm

Kumar, Srijan, Francesca Spezzano, V. S. Subrahmanian, and Christos Faloutsos, "Edge Weight Prediction in Weighted Signed Networks," *2016 IEEE 16th International Conference on Data Mining (ICDM)*, 2016.

McLeod, Gary, George Nacouzi, Paul Dreyer, Mel Eisman, Myron Hura, Krista Langeland, David Manheim, and Geoffrey Torrington, *Enhancing Space Resilience Through Non-Materiel Means*, RAND Corporation, RR-1067-AF, 2016. As of April 7, 2023:
https://www.rand.org/pubs/research_reports/RR1067.html

Merriam-Webster, "joint," dictionary entry, undated. As of June 13, 2023:
https://www.merriam-webster.com/dictionary/joint#dictionary-entry-2

NASA Space Science Data Coordinated Archive, "IDSCP 3-1," webpage, National Aeronautics and Space Administration, undated. As of July 2021:
https://nssdc.gsfc.nasa.gov/nmc/spacecraft/display.action?id=1967-066A

Norton, Keith, "Commercial SATCOM Remains Vital to Military Ops," *Defense One*, August 22, 2011.

Office of the Assistant Secretary of Defense for Homeland Defense and Global Security, *Space Domain Mission Assurance: A Resilience Taxonomy*, September 2015.

Phillips, Kathryn A., F. Reed Johnson, and Tara Maddala, "Measuring What People Value: A Comparison of 'Attitude' and 'Preference' Surveys," *Health Services Research*, Vol. 37, No. 6, December 2002.

Robins, Garry, *Doing Social Network Research: Network-Based Research Design for Social Scientists*, SAGE Publications, 2015.

U.S. Department of Defense Directive 3100.10, *Space Policy*, Office of the Under Secretary of Defense for Policy, August 30, 2022.

White House, *National Space Policy of the United States of America*, June 28, 2010. As of July 2021:

https://obamawhitehouse.archives.gov/sites/default/files/national_space_policy_6-28-10.pdf